Collins

5 Minute Maths Mastery

Chief editor: Zhou Jieying
Consultant: Fan Lianghuo

Year 4

CONTENTS

How to use this book 4

Test 1	1000 more or less 5
Test 2	Knowing numbers up to 10 000 (1) 6
Test 3	Knowing numbers up to 10 000 (2) 7
Test 4	Rounding numbers to the nearest 10, 100, 1000 8
Test 5	Roman numerals 9
Test 6	Addition with three-digit numbers (1) 10
Test 7	Addition with three-digit numbers (2) 11
Test 8	Subtraction with three-digit numbers (1) 12
Test 9	Subtraction with three-digit numbers (2) 13
Test 10	Addition and subtraction with three-digit numbers (1) 14
Test 11	Addition and subtraction with three-digit numbers (2) 15
Test 12	Addition and subtraction with three-digit numbers (3) 16
Test 13	Relationship between addition and subtraction (1) 17
Test 14	Relationship between addition and subtraction (2) 18
Test 15	Estimating additions and subtractions with four-digit numbers (1) 19
Test 16	Addition with four-digit numbers (1) 20
Test 17	Addition with four-digit numbers (2) 21
Test 18	Subtraction with four-digit numbers (1) 22
Test 19	Subtraction with four-digit numbers (2) 23
Test 20	Multiplication and division of 6 (1) 24
Test 21	Multiplication and division of 6 (2) 25
Test 22	Multiplication and division of 7 (1) 26
Test 23	Multiplication and division of 7 (2) 27
Test 24	Multiplication and division of 7 (3) 28
Test 25	Multiplication and division of 9 (1) 29
Test 26	Multiplication and division of 9 (2) 30
Test 27	Multiplication and division (1) 31
Test 28	Multiplication and division (2) 32
Test 29	5 threes plus 3 threes equals 8 threes 33
Test 30	5 threes minus 3 threes equals 2 threes (1) 34
Test 31	5 threes minus 3 threes equals 2 threes (2) 35
Test 32	Practice (1) – Multiples 36
Test 33	Practice (2) – Multiples 37
Test 34	Multiplying whole tens by a two-digit number (1) 38
Test 35	Multiplying whole tens by a two-digit number (2) 39
Test 36	Multiplying whole tens by a two-digit number (3) 40
Test 37	Multiplying a two-digit number by a two-digit number (1) 41
Test 38	Multiplying a two-digit number by a two-digit number (2) 42
Test 39	Multiplying a two-digit number by a two-digit number (3) 43
Test 40	Multiplication and division of two-digit numbers 44

Test 41	Multiplying a three-digit number by a one-digit number (1)	45
Test 42	Multiplying a three-digit number by a one-digit number (2)	46
Test 43	Dividing three-digit numbers by tens (1)	47
Test 44	Dividing three-digit numbers by tens (2)	48
Test 45	Dividing a three-digit number by a one-digit number (1)	49
Test 46	Dividing a three-digit number by a one-digit number (2)	50
Test 47	Column multiplication (1)	51
Test 48	Column multiplication (2)	52
Test 49	Relationship between multiplication and division (1)	53
Test 50	Relationship between multiplication and division (2)	54
Test 51	Fractions (1)	55
Test 52	Fractions (2)	56
Test 53	Addition of fractions	57
Test 54	Subtraction of fractions	58
Test 55	Addition and subtraction of fractions	59
Test 56	Practice with fractions (1)	60
Test 57	Practice with fractions (2)	61
Test 58	Decimals in life	62
Test 59	Understanding decimals (1)	63
Test 60	Understanding decimals (2)	64
Test 61	Understanding decimals (3)	65
Test 62	Understanding decimals (4)	66
Test 63	Understanding decimals (5)	67
Test 64	Understanding decimals (6)	68
Test 65	Comparing decimals (1)	69
Test 66	Comparing decimals (2)	70
Test 67	Practice (1)	71
Test 68	Practice (2)	72
Test 69	Acute angles and obtuse angles	73
Test 70	Triangles, quadrilaterals and pentagons	74
Test 71	Classification of triangles	75
Test 72	Line symmetry	76
Test 73	Areas of rectangles and squares (1)	77
Test 74	Areas of rectangles and squares (2)	78
Test 75	Areas of rectangles and squares (3)	79
Test 76	Converting between kilometres and metres	80
Test 77	Perimeters of rectangles and squares (1)	81
Test 78	Perimeters of rectangles and squares (2)	82
Test 79	Solving problems involving time and money	83
Test 80	Solving calculations in steps	84
Answers		85

HOW TO USE THIS BOOK

The best way to help children to achieve mastery in maths is to give them lots and lots of practice in the key facts and skills.

Written by maths mastery experts, this series will help children to become fluent in number facts, and help them to recall them quickly – both are essential to mastering maths.

This photocopiable resource saves time with ready-to-practise questions that comprehensively cover the number curriculum for Year 4. It contains 80 topic-based tests, each 5 minutes long, to help children build up their mathematical fluency.

Each test is divided into three Steps:

- **Step 1: Warm-up (1 minute)**
  This exercise helps children to revise maths they should already know and gives them preparation for Step 2.
- **Step 2: Rapid calculation ($2\frac{1}{2}$ minutes)**
  This exercise requires children to answer a set of questions focused on the topic area being tested.
- **Step 3: Challenge ($1\frac{1}{2}$ minutes)**
  This is a more testing exercise designed to stretch the mental abilities of children.

Some of the tests also include:

- a Tip to help children answer questions of a particular type.
- a Mind Gym puzzle – this is a further test of mental agility and is not included in the 5-minute time allocation.

Children should attempt to answer as many questions as possible in the time allowed at each Step. Answers are provided at the back of the book.

To help to measure progress, each test includes boxes for recording the date, the total score obtained and the total time taken.

ACKNOWLEDGEMENTS

The authors and publisher are grateful to the copyright holders for permission to use quoted materials and images.

All images are © HarperCollins*Publishers* Ltd and © Shutterstock.com

Every effort has been made to trace copyright holders and obtain their permission for the use of copyright material. The authors and publisher will gladly receive information enabling them to rectify any error or omission in subsequent editions. All facts are correct at time of going to press.

Published by Collins in association with East China Normal University Press

Collins
An imprint of HarperCollins*Publishers*
1 London Bridge Street
London SE1 9GF

ISBN: 978-0-00-831117-9

First published 2019

10 9 8 7 6 5 4 3 2 1

©HarperCollins*Publishers* Ltd. 2019,
©East China Normal University Press Ltd.,
©Zhou Jieying

All rights reserved. No part of this publication may be reproduced, stored in a retrieval system, or transmitted, in any form or by any means, electronic, mechanical, photocopying, recording or otherwise, without the prior permission of Collins.

British Library Cataloguing in Publication Data.

A CIP record of this book is available from the British Library.

Publisher: Fiona McGlade
Consultant: Fan Lianghuo
Authors: Zhou Jieying, Chen Weihua and Xu Jing
Editors: Ni Ming and Xu Huiping
Contributor: Paul Hodge
Project Management and Editorial: Richard Toms, Lauren Murray and Marie Taylor
Cover Design: Sarah Duxbury
Inside Concept Design: Paul Oates and Ian Wrigley
Layout: Jouve India Private Limited

Printed by CPI Group (UK) Ltd, Croydon

MIX
Paper from responsible source
FSC® C007454

This book is produced from independently certified FSC™ paper to ensure responsible forest management.

For more information visit:
www.harpercollins.co.uk/green

1000 more or less

Date: _____
Day of Week: _____

STEP 1 — Warm-up (1 min)

Answer these.

23 + 10 = ☐   428 − 10 = ☐   504 + 10 = ☐   667 − 10 = ☐

346 + 100 = ☐   473 − 100 = ☐   88 + 100 = ☐   299 − 100 = ☐

970 + 100 = ☐   901 − 100 = ☐

STEP 2 — Rapid calculation (2.5 min)

Answer these.

30 + 1000 = ☐   1100 − 1000 = ☐   90 + 1000 = ☐

2300 − 1000 = ☐   1400 + 1000 = ☐   4700 − 3700 = ☐

4650 + 1000 = ☐   8460 − 1000 = ☐   2300 − 1000 = ☐

9309 − 1000 = ☐   5781 − 1000 = ☐   5077 + 1000 = ☐

4356 + 1000 = ☐   1400 + 1000 = ☐   9720 + 1000 = ☐

STEP 3 — Challenge (1.5 min)

Fill in the missing numbers.

740 + ☐ = 840   6352 + ☐ = 7352

5871 − ☐ = 5771   4867 − ☐ = 4857

3499 + ☐ = 4499   3038 + ☐ = 3138

9548 − ☐ = 8548   4895 − ☐ = 4885

Time spent: ____ min ____ sec. Total: ____ out of 33

2 Knowing numbers up to 10 000 (1)

Date: _____
Day of Week: _____

STEP 1 — Warm-up (1 min)

Write each of these numbers in words on a separate sheet of paper.

7326	3821	8230	9526
9662	5600	3702	8714

STEP 2 — Rapid calculation (2.5 min)

Fill in the missing numbers. The first one has been done for you.

6428 = **6000** + **400** + **20** + **8**

1560 = ☐ + ☐ + ☐ + ☐

3395 = ☐ + ☐ + ☐ + ☐

4586 = ☐ + ☐ + ☐ + ☐

3499 = ☐ + ☐ + ☐ + ☐

8550 = ☐ + ☐ + ☐ + ☐

7788 = ☐ + ☐ + ☐ + ☐

2800 = ☐ + ☐ + ☐ + ☐

9065 = ☐ + ☐ + ☐ + ☐

5080 = ☐ + ☐ + ☐ + ☐

7086 = ☐ + ☐ + ☐ + ☐

6034 = ☐ + ☐ + ☐ + ☐

STEP 3 — Challenge (1.5 min)

Write these numbers in digits.

2 thousands and 1 hundred	
9 thousands, 8 hundreds and 5 tens	
4 thousands, 6 hundreds, 2 tens and 6 ones	
7 thousands and 3 ones	
5 thousands and 3 tens	
4 thousands, 1 hundred and 6 ones	
6 thousands, 8 hundreds and 2 tens	
9 thousands and 4 hundreds	

Time spent: _____ min _____ sec. Total: _____ out of 27

Knowing numbers up to 10 000 (2)

Date: _____
Day of Week: _____

STEP 1 (1 min) Warm-up

Write these numbers in figures.

four thousand, three hundred and eighty-five ☐

two thousand, four hundred and ninety-seven ☐

nine thousand, three hundred ☐

three thousand, nine hundred ☐

six thousand and fifty-four ☐

eight thousand and fourteen ☐

seven thousand, six hundred and ninety ☐

five thousand and five ☐

STEP 2 (2.5 min) Rapid calculation

Answer these.

4000 + 900 + 80 = ☐   5000 + 300 + 20 + 8 = ☐   6000 + 300 + 20 = ☐

7000 + 80 + 5 = ☐   3000 + 30 = ☐   7000 + 800 + 90 + 1 = ☐

2000 + 90 + 5 = ☐   3000 + 900 + 50 + 7 = ☐   4000 + 800 + 6 = ☐

6000 + 800 + 50 + 3 = ☐   5000 + 400 + 50 + 9 = ☐   2000 + 5 = ☐

STEP 3 (1.5 min) Challenge

TIP *You can use your fingers to remember the place order of numbers up to 10 000. Spread your left hand with palm towards you, and write the places on the fingers: write ones on your little finger, tens on the fourth finger, hundreds on the middle finger, thousands on the index finger and ten thousands on the thumb. Now you have the order on a hand.*

1. Find the nearest whole tens. The first one has been done for you.

   3940 ← 3945 → 3950   ☐ ← 8992 → ☐   ☐ ← 5777 → ☐

2. Find the nearest whole hundreds.

   ☐ ← 6390 → ☐   ☐ ← 2784 → ☐   ☐ ← 9750 → ☐

3. Find the nearest whole thousands.

   ☐ ← 4028 → ☐   ☐ ← 6835 → ☐   ☐ ← 8499 → ☐

Time spent: _____ min _____ sec. Total: _____ out of 28

4 Rounding numbers to the nearest 10, 100, 1000

Date: _____
Day of Week: _____

STEP 1 (1 min) Warm-up

1. Find the nearest whole tens.

   ☐ ← 3901 → ☐   ☐ ← 5831 → ☐   ☐ ← 4405 → ☐

2. Find the nearest whole hundreds.

   ☐ ← 3290 → ☐   ☐ ← 4582 → ☐   ☐ ← 5349 → ☐

3. Find the nearest whole thousands.

   ☐ ← 9804 → ☐   ☐ ← 8313 → ☐   ☐ ← 2500 → ☐

STEP 2 (2.5 min) Rapid calculation

Start the timer

Fill in the missing numbers.

Number	4972	3021	2389	8951	9314
Nearest whole tens					

Number	3401	4718	7382	6478	5639
Nearest whole hundreds					

Number	4890	3056	2795	4200	5841
Nearest whole thousands					

STEP 3 (1.5 min) Challenge

Start the timer

Find the pattern and fill in the missing numbers for each sequence.

2150, 2200, 2250, ☐, ☐   7685, 7585, ☐, ☐, 7285

9753, 7753, ☐, ☐, 1753   3206, ☐, 3306, ☐, 3406

115, 315, ☐, ☐, 915   5100, 6100, ☐, ☐, 9100

3809, 3810, 3811, ☐, ☐   4020, 4130, 4240, ☐, ☐

Time spent: _____ min _____ sec. Total: _____ out of 32

Roman numerals 5

STEP 1 (1 min) **Warm-up**

Write the value of each Roman numeral.

I = ☐   V = ☐   X = ☐   L = ☐

C = ☐   D = ☐   M = ☐

STEP 2 (2.5 min) **Rapid calculation**

Write each Roman numeral as an Arabic numeral.

VIII = ☐   IV = ☐   XII = ☐

XV = ☐   XLIII = ☐   LXVI = ☐

IX = ☐   LXXXII = ☐   XXXVII = ☐

STEP 3 (1.5 min) **Challenge**

Fill in the boxes using >, < or =.

VII ☐ 6   XIX ☐ 19   XXVIII ☐ 29

XXXIII ☐ 32   LXXIV ☐ 75   XCVI ☐ 96

Time spent: _____ min _____ sec. Total: _____ out of 22

6 Addition with three-digit numbers (1)

Date: _____
Day of Week: _____

STEP 1 Warm-up

Answer these.

400 + 300 = ☐   250 + 500 = ☐   700 + 290 = ☐   180 + 320 = ☐

680 + 240 = ☐   270 + 440 = ☐   234 + 400 = ☐   500 + 179 = ☐

STEP 2 Rapid calculation

Answer these.

480 + 235 = ☐   250 + 480 = ☐   690 + 246 = ☐

553 + 372 = ☐   364 + 324 = ☐   437 + 252 = ☐

625 + 253 = ☐   177 + 422 = ☐   851 + 147 = ☐

274 + 323 = ☐   145 + 641 = ☐   157 + 342 = ☐

444 + 238 = ☐   545 + 383 = ☐   627 + 282 = ☐

723 + 247 = ☐   457 + 345 = ☐   207 + 579 = ☐

776 + 188 = ☐   856 + 197 = ☐

STEP 3 Challenge

Answer these.

456 + 795 = ☐   567 + 687 = ☐   777 + 576 = ☐   429 + 567 = ☐

549 + 676 = ☐   294 + 998 = ☐   848 + 867 = ☐   968 + 966 = ☐

Time spent: _____ min _____ sec. Total: _____ out of 36

Date: _____
Day of Week: _____

Addition with three-digit numbers (2)

STEP 1 — Warm-up (1 min)

Answer these.

600 + 350 = ☐   500 + 420 = ☐   300 + 470 = ☐   660 + 240 = ☐

710 + 190 = ☐   550 + 370 = ☐   454 + 320 = ☐   677 + 220 = ☐

STEP 2 — Rapid calculation (2.5 min)

Answer these.

455 + 342 = ☐   546 + 333 = ☐   767 + 133 = ☐

466 + 423 = ☐   436 + 156 = ☐   566 + 235 = ☐

546 + 345 = ☐   234 + 348 = ☐   473 + 474 = ☐

643 + 358 = ☐   347 + 345 = ☐   743 + 145 = ☐

436 + 354 = ☐   736 + 136 = ☐   362 + 479 = ☐

784 + 197 = ☐   535 + 396 = ☐   437 + 473 = ☐

376 + 588 = ☐   874 + 167 = ☐

STEP 3 — Challenge (1.5 min)

Answer these.

436 + 678 = ☐   567 + 676 = ☐   547 + 766 = ☐   657 + 634 = ☐

687 + 796 = ☐   786 + 768 = ☐   787 + 838 = ☐   974 + 978 = ☐

Time spent: _____ min _____ sec. Total: _____ out of 36

8 Subtraction with three-digit numbers (1)

Date: _____
Day of Week: _____

STEP 1 Warm-up (1 min)

Answer these.

63 – 28 = ☐   54 – 28 = ☐   72 – 27 = ☐   65 – 29 = ☐

77 – 58 = ☐   65 – 26 = ☐   82 – 53 = ☐   78 – 39 = ☐

91 – 46 = ☐   86 – 67 = ☐

STEP 2 Rapid calculation (2.5 min)

Answer these.

455 – 233 = ☐   566 – 444 = ☐   435 – 122 = ☐   656 – 413 = ☐

545 – 114 = ☐   734 – 231 = ☐   748 – 326 = ☐   976 – 754 = ☐

576 – 238 = ☐   546 – 227 = ☐   876 – 649 = ☐   663 – 248 = ☐

746 – 288 = ☐   567 – 439 = ☐   677 – 398 = ☐   954 – 375 = ☐

323 – 182 = ☐   711 – 526 = ☐

STEP 3 Challenge (1.5 min)

Fill in the boxes with >, < or =.

565 – 386 ☐ 180   725 – 347 ☐ 377   676 – 188 ☐ 488

456 – 287 ☐ 170   911 – 465 ☐ 445   823 – 546 ☐ 277

936 – 778 ☐ 157   655 – 266 ☐ 390   564 – 289 ☐ 274

Time spent: _____ min _____ sec. Total: _____ out of 37

Subtraction with three-digit numbers (2)

STEP 1 — Warm-up

Answer these.

47 − 19 =   64 − 56 =   75 − 38 =   56 − 28 =

45 − 27 =   93 − 45 =   76 − 57 =   94 − 55 =

86 − 68 =   67 − 28 =

STEP 2 — Rapid calculation

Answer these.

548 − 439 =   446 − 228 =   347 − 128 =   773 − 556 =

674 − 185 =   952 − 437 =   882 − 666 =   945 − 618 =

656 − 264 =   455 − 178 =   547 − 189 =   488 − 399 =

567 − 278 =   923 − 655 =   852 − 475 =   644 − 266 =

245 − 111 =   339 − 127 =

STEP 3 — Challenge

Fill in the missing numbers.

343 + ☐ = 527   ☐ + 246 = 624

642 + ☐ = 921   ☐ + 358 = 852

458 + ☐ = 826   ☐ + 275 = 543

635 + ☐ = 922   ☐ + 738 = 927

10 Addition and subtraction with three-digit numbers (1)

Date: _____
Day of Week: _____

STEP 1 (1 min) Warm-up

Answer these.

410 − 200 = ☐   220 + 100 = ☐   650 + 300 = ☐   520 + 400 = ☐

290 + 500 = ☐   780 + 200 = ☐   390 − 100 = ☐   420 − 300 = ☐

640 − 400 = ☐   980 − 500 = ☐

STEP 2 (2.5 min) Rapid calculation

Answer these.

417 − 300 = ☐   559 + 300 = ☐   620 + 209 = ☐   360 − 150 = ☐

865 − 260 = ☐   785 − 280 = ☐   145 + 150 = ☐   450 + 160 = ☐

125 + 270 = ☐   470 − 330 = ☐   572 − 170 = ☐   656 + 240 = ☐

712 + 180 = ☐   220 + 452 = ☐   1000 − 500 = ☐   624 − 448 = ☐

STEP 3 (1.5 min) Challenge

Fill in the missing numbers.

280 + ☐ = 490   220 + ☐ = 795   345 + ☐ = 565

306 − ☐ = 140   ☐ − 270 = 400   540 + ☐ = 880

895 − ☐ = 350   ☐ − 320 = 550   ☐ + 120 = 695

Time spent: _____ min _____ sec. Total: _____ out of 35

Addition and subtraction with three-digit numbers (2)

STEP 1 — Warm-up

Answer these.

460 + 620 = ☐   260 + 200 = ☐   1000 − 300 = ☐   340 + 200 = ☐

570 − 300 = ☐   630 + 300 = ☐   720 − 400 = ☐   410 + 400 = ☐

300 + 490 = ☐   780 − 500 = ☐

STEP 2 — Rapid calculation

Fill in the missing numbers.

195 + 700 = ☐   810 − 350 = ☐   204 + 390 = ☐

532 − 202 = ☐   731 − 231 = ☐   810 − 300 = ☐

416 + 380 = ☐   845 − 205 = ☐   540 + 450 = ☐

390 − 270 = ☐   104 + ☐ = 309   643 − ☐ = 230

850 − ☐ = 520   ☐ − 452 = 330   ☐ − 159 = 430

STEP 3 — Challenge

Fill in the missing numbers.

456 − 120 = ☐   248 + 320 = ☐   973 − 250 = ☐

656 − 120 = ☐   248 + 340 = ☐   ☐ − 250 = 423

☐ − 120 = 736   248 + ☐ = 608   373 − 250 = ☐

12 Addition and subtraction with three-digit numbers (3)

Date: _____
Day of Week: _____

STEP 1 Warm-up

Answer these.

42 − 24 = ☐   58 − 39 = ☐   46 + 28 = ☐   63 − 47 = ☐

81 − 56 = ☐   24 + 67 = ☐   56 + 16 = ☐   72 − 63 = ☐

71 − 63 + 23 = ☐   45 + 28 + 19 = ☐

STEP 2 Rapid calculation

Answer these.

234 + 432 = ☐   697 − 284 = ☐   359 − 148 = ☐   456 + 432 = ☐

562 + 126 = ☐   621 − 508 = ☐   956 − 529 = ☐   495 − 268 = ☐

582 − 245 = ☐   444 + 446 = ☐   827 − 608 = ☐   548 + 223 = ☐

426 + 238 = ☐   745 − 319 = ☐   358 + 319 = ☐   646 + 219 = ☐

754 − 647 = ☐   257 + 436 = ☐   508 + 475 = ☐   396 − 147 = ☐

STEP 3 Challenge

Fill in the missing numbers.

☐ + 16 = 482   746 − ☐ = 64   ☐ + 21 = 290

266 − ☐ = 18   98 + ☐ = 788   47 + ☐ = 853

☐ + 29 = 346   ☐ + 18 = 267   91 + ☐ = 702

Time spent: _____ min _____ sec. Total: _____ out of 39

Relationship between addition and subtraction (1)

Date: _____
Day of Week: _____

STEP 1 — Warm-up

Answer these.

28 + 52 = ☐   80 − 52 = ☐   80 − 28 = ☐   91 − 37 = ☐

54 + 37 = ☐   91 − 54 = ☐   260 + 530 = ☐   790 − 260 = ☐

790 − 530 = ☐   1200 − 400 = ☐

STEP 2 — Rapid calculation

TIP *Remember the inverse relationship between addition and subtraction.*

Fill in the missing numbers.

☐ + 12 = 60   72 + ☐ = 100   ☐ + 45 = 155

347 − ☐ = 227   1000 − ☐ = 230   125 + ☐ = 200

☐ − 39 = 46   ☐ − 81 = 81   93 + ☐ = 153

☐ − 430 = 250   ☐ − 310 = 210   112 + ☐ = 160

☐ + 70 = 320   ☐ − 350 = 650   729 − ☐ = 420

STEP 3 — Challenge

Fill in the missing numbers.

☐ − 64 = 36   372 − ☐ = 122   ☐ + 180 = 340

☐ − 50 = 5 × 24   402 + ☐ − 224 = 200   ☐ + 75 = 5 × 48

☐ − 650 = 200 + 150   81 − ☐ = 119 − 59

14 Relationship between addition and subtraction (2)

Date: _____
Day of Week: _____

STEP 1 Warm-up

Answer these.

840 + 130 = ☐   46 + 94 = ☐   450 + 350 = ☐   3200 + 480 = ☐

970 − 840 = ☐   140 − 46 = ☐   800 − 450 = ☐   3680 − 480 = ☐

970 − 130 = ☐   140 − 94 = ☐

STEP 2 Rapid calculation

Fill in the missing numbers.

☐ + 27 = 80   170 + ☐ = 240   ☐ − 140 = 800

1000 − ☐ = 270   793 − ☐ = 83   260 + ☐ = 300

☐ − 130 = 260   ☐ − 160 = 160   ☐ − 620 = 380

34 + ☐ = 93   374 − ☐ = 143   350 + ☐ = 750

176 − ☐ = 46   ☐ − 26 = 150   790 − ☐ = 200

STEP 3 Challenge

Fill in the missing numbers.

☐ − 24 = 66   ☐ + 518 = 938   ☐ − 250 = 15 × 24

☐ − 56 = 79 + 21   672 − ☐ = 270   ☐ + 800 = 25 × 48

☐ + 320 = 480 + 100   380 − ☐ = 490 − 150

Time spent: ____ min ____ sec. Total: ____ out of 33

Estimating additions and subtractions with four-digit numbers (1)

Date: _____
Day of Week: _____

STEP 1 — Warm-up (1 min)

Write these numbers to the nearest hundreds.

4582	___	3367	___	5876	___	9050	___	6911	___
8436	___	5555	___	1735	___	4382	___	6450	___

STEP 2 — Rapid calculation (2.5 min)

Estimate and then calculate.

Question	Estimate	Calculated answer
4374 + 2318 =		
1458 + 7136 =		
5436 + 2357 =		
4835 + 3197 =		

Question	Estimate	Calculated answer
5469 − 2587 =		
7437 − 3368 =		
8435 − 2157 =		
4354 − 1177 =		

STEP 3 — Challenge (1.5 min)

Estimate and then calculate.

Question	Estimate	Calculated answer
7353 − 2578 =		
5478 + 3784 =		
9467 − 6688 =		

Question	Estimate	Calculated answer
3285 + 4986 =		
8844 − 3967 =		
4378 + 4884 =		

Time spent: ___ min ___ sec. Total: ___ out of 24

16 Addition with four-digit numbers (1)

Date: _____
Day of Week: _____

STEP 1 (1 min) Warm-up

Answer these.

300 + 200 =   500 + 4000 =   2000 + 4000 =

800 + 500 =   3000 + 900 =   4000 + 5000 =

3000 + 6000 =   1000 + 7000 =   2000 + 7000 =

STEP 2 (2.5 min) Rapid calculation

Answer these.

3200 + 1700 =   2800 + 1100 =   1300 + 7100 =

6200 + 2700 =   5500 + 4200 =   4300 + 2500 =

5300 + 2400 =   4500 + 5400 =   3100 + 2800 =

3700 + 2100 =   1200 + 2400 =   6500 + 2300 =

3200 + 3600 =   7200 + 2200 =   6300 + 2400 =

STEP 3 (1.5 min) Challenge

Answer these.

1203 + 2203 =   2304 + 3440 =   4102 + 2475 =

3102 + 2457 =   3540 + 2321 =   4362 + 1537 =

5903 + 3024 =   1004 + 3024 =

Time spent: _____ min _____ sec. Total: _____ out of 32

Addition with four-digit numbers (2)

Date: _____
Day of Week: _____

STEP 1 — Warm-up (1 min)

Answer these.

230 + 340 = ☐   174 + 420 = ☐   540 + 320 = ☐   810 + 109 = ☐

630 + 207 = ☐   450 + 210 = ☐   350 + 230 = ☐   325 + 250 = ☐

428 + 130 = ☐   375 + 200 = ☐

STEP 2 — Rapid calculation (2.5 min)

TIP: *When doing addition with multi-digit numbers, line up the same digit places and start from the ones place. If the sum is ten or over, carry 1 to the column on the left.*

Answer these.

2400 + 1200 = ☐   801 + 2000 = ☐   4800 + 2100 = ☐

5108 + 2030 = ☐   3500 + 6400 = ☐   2300 + 4500 = ☐

3904 + 2002 = ☐   1700 + 3094 = ☐   4974 + 3811 = ☐

1862 + 2103 = ☐   3593 + 4614 = ☐   5222 + 3333 = ☐

STEP 3 — Challenge (1.5 min)

Answer these.

3905 + 5996 = ☐   2857 + 4926 = ☐   2683 + 4597 = ☐

4728 + 2965 = ☐   3920 + 4682 = ☐   4982 + 1636 = ☐

3978 + 4926 = ☐   2936 + 4385 = ☐

Time spent: _____ min _____ sec. Total: _____ out of 30

18 Subtraction with four-digit numbers (1)

Date: _____
Day of Week: _____

STEP 1 — Warm-up (1 min)

Answer these.

400 – 200 = ☐   900 – 50 = ☐   800 – 600 = ☐   600 – 300 = ☐

700 – 200 = ☐   1100 – 500 = ☐   500 – 400 = ☐   700 – 400 = ☐

400 – 100 = ☐   900 – 300 = ☐

STEP 2 — Rapid calculation (2.5 min)

Answer these.

4200 – 2100 = ☐   8700 – 5200 = ☐   7400 – 3300 = ☐

7600 – 4300 = ☐   6700 – 5200 = ☐   7300 – 4300 = ☐

9500 – 4100 = ☐   5800 – 2600 = ☐   8400 – 2400 = ☐

4800 – 2400 = ☐   5800 – 3200 = ☐   6900 – 3600 = ☐

6600 – 4400 = ☐   7900 – 3500 = ☐   5900 – 4100 = ☐

STEP 3 — Challenge (1.5 min)

Answer these.

5302 – 2201 = ☐   7450 – 2320 = ☐   5479 – 2361 = ☐

4090 – 2030 = ☐   7489 – 3265 = ☐   6590 – 4560 = ☐

4807 – 2405 = ☐   5784 – 2561 = ☐

Time spent: ____ min ____ sec. Total: ____ out of 33

Subtraction with four-digit numbers (2)

STEP 1 Warm-up

Answer these.

4000 – 2000 =   9000 – 4000 =

8000 – 3000 =   4000 – 100 =

7500 – 4000 =   5000 – 1000 =

8000 – 5000 =   9000 – 2000 =

3500 – 2000 =   8600 – 5000 =

STEP 2 Rapid calculation

Calculate step by step.

5427 – 3905 =   7392 – 3948 =   8320 – 4732 =

8301 – 4983 =   6835 – 5941 =   7431 – 6392 =

5930 – 3867 =   7395 – 5499 =   6201 – 4953 =

6544 – 4207 =   9001 – 6549 =   5455 – 3216 =

STEP 3 Challenge

Answer these.

3200 – 1200 =   4309 – 2204 =   9487 – 4355 =

7579 – 5260 =   4500 – 2300 =   8907 – 3405 =

5498 – 3060 =   4990 – 3650 =

20 Multiplication and division of 6 (1)

Date: _____
Day of Week: _____

STEP 1 Warm-up

Answer these.

0 ÷ 5 = ☐   6 × 0 = ☐   64 ÷ 8 = ☐   55 + 24 = ☐

77 ÷ 7 = ☐   36 − 26 = ☐   18 ÷ 9 = ☐   1 × 6 = ☐

36 ÷ 3 = ☐   12 × 5 = ☐   29 + 34 = ☐   6 × 8 = ☐

STEP 2 Rapid calculation

Answer these.

2 × 6 = ☐   8 × 6 = ☐   6 × 3 = ☐   11 × 6 = ☐

6 × 5 = ☐   6 × 12 = ☐   6 × 7 = ☐   9 × 6 = ☐

24 ÷ 6 = ☐   30 ÷ 6 = ☐   18 ÷ 6 = ☐   54 ÷ 6 = ☐

48 ÷ 6 = ☐   10 × 6 = ☐   6 ÷ 6 = ☐   6 × 9 = ☐

66 ÷ 6 = ☐   6 ÷ 1 = ☐   6 × 6 = ☐   60 ÷ 6 = ☐

STEP 3 Challenge

Answer these.

6 × 7 + 3 = ☐   9 × 6 + 6 = ☐   6 × 10 + 14 = ☐

8 × 6 + 11 = ☐   12 × 6 + 18 = ☐   36 ÷ 6 + 5 = ☐

30 ÷ 10 + 16 = ☐   18 ÷ 6 + 72 = ☐   72 ÷ 6 + 81 = ☐

Time spent: _____ min _____ sec. Total: _____ out of 41

Multiplication and division of 6 (2) — 21

STEP 1 — Warm-up (1 min)

Answer these.

12 ÷ 3 = ☐   42 + 36 = ☐   45 ÷ 9 = ☐   36 − 29 = ☐

8 × 11 = ☐   54 − 36 = ☐   18 ÷ 6 = ☐   8 × 9 = ☐

18 ÷ 2 = ☐   12 × 6 = ☐   26 + 57 = ☐   48 ÷ 4 = ☐

STEP 2 — Rapid calculation (2.5 min)

Answer these.

6 × 10 = ☐   42 ÷ 6 = ☐   66 ÷ 6 = ☐   9 × 6 = ☐

3 × 6 = ☐   30 ÷ 5 = ☐   7 × 6 = ☐   6 ÷ 6 = ☐

12 ÷ 2 = ☐   48 ÷ 8 = ☐   18 ÷ 3 = ☐   54 ÷ 6 = ☐

72 ÷ 6 = ☐   60 ÷ 6 = ☐   0 ÷ 6 = ☐   6 × 0 = ☐

42 ÷ 7 = ☐   36 ÷ 6 = ☐   0 ÷ 3 = ☐   24 ÷ 6 = ☐

STEP 3 — Challenge (1.5 min)

Fill in the missing numbers.

☐ ÷ 6 = 3   36 ÷ ☐ = 6   ☐ × 7 = 42

☐ ÷ 4 = 6   ☐ × 4 = 24   ☐ ÷ 9 = 6

8 × ☐ = 12 × 6   18 ÷ ☐ = 9 ÷ 3   ☐ ÷ 6 = 33 ÷ 3

Time spent: ___ min ___ sec. Total: ___ out of 41

22 Multiplication and division of 7 (1)

Date: _____
Day of Week: _____

STEP 1 Warm-up

Answer these.

4 × 8 = ☐   26 + 18 = ☐   72 ÷ 9 = ☐   56 − 28 = ☐

45 ÷ 5 = ☐   72 − 28 = ☐   5 × 6 = ☐   34 + 29 = ☐

100 − 46 = ☐   4 × 12 = ☐   62 + 9 = ☐   12 × 2 = ☐

STEP 2 Rapid calculation

Answer these.

7 × 2 = ☐   4 × 7 = ☐   8 × 7 = ☐   5 × 7 = ☐

7 × 6 = ☐   2 × 7 = ☐   7 × 3 = ☐   7 × 4 = ☐

7 × 5 = ☐   7 × 12 = ☐   7 × 8 = ☐   7 × 9 = ☐

1 × 7 = ☐   10 × 7 = ☐   3 × 7 = ☐   0 × 7 = ☐

7 × 11 = ☐   6 × 7 = ☐   12 × 7 = ☐   7 × 7 = ☐

STEP 3 Challenge

Answer these.

21 ÷ 7 = ☐   35 ÷ 7 = ☐   49 ÷ 7 = ☐

14 ÷ 7 = ☐   56 ÷ 7 = ☐   77 ÷ 7 = ☐

84 ÷ 7 = ☐   42 ÷ 7 = ☐   63 ÷ 7 = ☐

Time spent: _____ min _____ sec. Total: _____ out of 41

Multiplication and division of 7 (2)

Date: _____
Day of Week: _____

STEP 1 — Warm-up (1 min)

Answer these.

5 × 6 = ☐   46 − 27 = ☐   24 ÷ 3 = ☐   60 − 27 = ☐

40 ÷ 5 = ☐   75 − 36 = ☐   5 × 9 = ☐   4 × 9 = ☐

10 × 6 = ☐   4 × 7 = ☐   9 × 5 = ☐   12 × 4 = ☐

STEP 2 — Rapid calculation (2.5 min)

Answer these.

3 × 7 = ☐   7 × 5 = ☐   7 × 7 = ☐   6 × 7 = ☐

7 × 11 = ☐   0 × 7 = ☐   7 × 9 = ☐   12 × 7 = ☐

49 ÷ 7 = ☐   35 ÷ 5 = ☐   56 ÷ 7 = ☐   21 ÷ 3 = ☐

77 ÷ 7 = ☐   70 ÷ 7 = ☐   63 ÷ 7 = ☐   0 ÷ 7 = ☐

70 ÷ 10 = ☐   42 ÷ 6 = ☐   63 ÷ 9 = ☐   84 ÷ 7 = ☐

STEP 3 — Challenge (1.5 min)

Answer these.

35 ÷ 7 + 5 = ☐   49 − 28 ÷ 7 = ☐

77 ÷ 7 + 16 = ☐   4 × 7 + 19 = ☐

5 × 7 − 12 = ☐   49 − 7 ÷ 7 = ☐

3 × 7 + 48 = ☐   12 × 7 − 43 = ☐

70 + 0 ÷ 7 = ☐

Mind Gym

◎ + ◎ = ☆ + ☆ + ☆ + ☆

△ + △ = ◎ + ◎ + ◎

1. ◎ = ☐ × ☆

   3 × ◎ = ☐ × ☆

2. 2 × △ = ☐ × ☆

   △ = ☐ × ☆

Time spent: _____ min _____ sec. Total: _____ out of 41

24 Multiplication and division of 7 (3)

Date: _____
Day of Week: _____

STEP 1 Warm-up

Answer these.

6 × 8 =   38 + 27 =   20 ÷ 4 =   65 − 56 =

45 ÷ 5 =   56 − 36 =   2 × 9 =   8 × 9 =

60 ÷ 5 =   5 × 7 =   28 + 44 =   11 × 4 =

STEP 2 Rapid calculation

Answer these.

7 × 7 =   7 × 6 =   7 × 9 =   5 × 7 =

7 × 0 =   11 × 7 =   14 ÷ 7 =   12 × 7 =

42 ÷ 7 =   49 ÷ 7 =   8 × 7 =   21 ÷ 7 =

0 ÷ 7 =   70 ÷ 10 =   56 ÷ 7 =   77 ÷ 7 =

42 ÷ 6 =   63 ÷ 9 =   84 ÷ 7 =   7 ÷ 1 =

STEP 3 Challenge

Fill in the missing numbers.

☐ ÷ 7 = 4   49 ÷ ☐ = 7   ☐ ÷ 8 = 12

☐ × 6 = 66   ☐ × 7 = 21   ☐ ÷ 7 = 7

84 ÷ ☐ = 12   ☐ × 11 = 77   ☐ ÷ 7 = 0

Time spent: ___ min ___ sec. Total: ___ out of 41

Multiplication and division of 9 (1)

STEP 1 — Warm-up (1 min)

Answer these.

3 × 11 = ☐   63 − 45 = ☐   10 × 0 = ☐   39 + 42 = ☐

7 × 12 = ☐   12 × 6 = ☐   77 ÷ 7 = ☐   2 × 11 = ☐

90 − 69 = ☐   0 ÷ 6 = ☐   42 − 14 = ☐   36 ÷ 3 = ☐

STEP 2 — Rapid calculation (2.5 min)

Answer these.

2 × 9 = ☐   5 × 9 = ☐   9 × 9 = ☐   9 × 10 = ☐

9 × 6 = ☐   11 × 9 = ☐   7 × 9 = ☐   9 × 8 = ☐

4 × 9 = ☐   9 × 12 = ☐   0 × 9 = ☐   54 ÷ 9 = ☐

18 ÷ 9 = ☐   108 ÷ 9 = ☐   81 ÷ 9 = ☐   99 ÷ 9 = ☐

63 ÷ 9 = ☐   27 ÷ 9 = ☐   45 ÷ 9 = ☐   90 ÷ 9 = ☐

STEP 3 — Challenge (1.5 min)

Fill in the missing numbers.

☐ ÷ 9 = 3   36 ÷ ☐ = 4   ☐ × 9 = 45

☐ ÷ 9 = 7   ☐ × 9 = 81   ☐ ÷ 9 = 10

9 × ☐ = 6 × 6   81 ÷ ☐ = 27 ÷ 3   ☐ × 9 = 108 ÷ 2

Time spent: _____ min _____ sec. Total: _____ out of 41

26 Multiplication and division of 9 (2)

Date: _____
Day of Week: _____

STEP 1 Warm-up

Answer these.

6 × 6 = ☐   33 ÷ 3 = ☐   28 + 16 = ☐   8 × 10 = ☐

90 − 9 = ☐   11 × 9 = ☐   10 ÷ 2 = ☐   12 × 7 = ☐

54 − 15 = ☐   27 ÷ 3 = ☐   72 − 28 = ☐   45 ÷ 5 = ☐

STEP 2 Rapid calculation

Answer these.

0 × 9 = ☐   9 × 5 = ☐   10 × 9 = ☐   36 ÷ 9 = ☐

6 × 9 = ☐   27 ÷ 9 = ☐   9 × 9 = ☐   9 ÷ 9 = ☐

63 ÷ 9 = ☐   9 × 4 = ☐   45 ÷ 9 = ☐   90 ÷ 9 = ☐

18 ÷ 2 = ☐   99 ÷ 9 = ☐   81 ÷ 9 = ☐   0 ÷ 9 = ☐

63 ÷ 7 = ☐   7 × 9 = ☐   54 ÷ 6 = ☐   9 × 12 = ☐

STEP 3 Challenge

Fill in the boxes with >, < or =.

9 × 7 ☐ 7 × 9   9 − 9 ☐ 9 ÷ 9   9 × 1 ☐ 9 ÷ 1

0 × 9 ☐ 0 + 9   36 ÷ 9 ☐ 36 ÷ 6   4 × 9 ☐ 9 × 5

9 × 9 ☐ 3 × 6   72 ÷ 8 ☐ 3 × 3   45 ÷ 9 ☐ 54 ÷ 9

Time spent: _____ min _____ sec. Total: _____ out of 41

Multiplication and division (1) — 27

STEP 1 Warm-up (1 min)

Answer these.

79 × 2 =
270 ÷ 9 =
84 ÷ 6 =

160 ÷ 2 =
56 ÷ 4 =
34 × 4 =

3 × 320 =
120 × 5 =
2 × 35 =

STEP 2 Rapid calculation (2.5 min)

Answer these.

72 × 4 =
208 ÷ 4 =
29 × 6 =
65 × 5 =
630 ÷ 9 =

38 × 5 =
210 ÷ 2 =
330 ÷ 3 =
84 ÷ 3 =
7 × 62 =

6 × 44 =
96 ÷ 6 =
520 ÷ 4 =
128 ÷ 4 =
92 ÷ 4 =

STEP 3 Challenge (1.5 min)

Fill in the missing numbers.

13 × 7 =
784 ÷ 7 =
21 × ☐ = 189

120 × 5 =
729 ÷ 9 =
910 ÷ ☐ = 130

56 × 11 =
252 ÷ 6 =
248 ÷ ☐ = 31

Time spent: ___ min ___ sec. Total: ___ out of 33

28 Multiplication and division (2)

STEP 1 Warm-up (1 min)

Answer these.

34 × 2 =
124 × 3 =
23 × 9 =

104 ÷ 4 =
180 ÷ 5 =
500 ÷ 4 =

530 ÷ 5 =
824 × 0 =
900 ÷ 6 =

STEP 2 Rapid calculation (2.5 min)

Answer these.

315 ÷ 5 =
252 ÷ 7 =
990 ÷ 5 =
247 × 3 =
120 × 8 =

879 ÷ 3 =
624 ÷ 6 =
126 × 5 =
3 × 503 =
270 ÷ 6 =

568 ÷ 8 =
864 ÷ 4 =
521 × 4 =
8 × 210 =
161 ÷ 7 =

STEP 3 Challenge (1.5 min)

Answer these.

240 ÷ 5 =
106 × 8 =
55 × 15 =
365 ÷ 5 =

350 ÷ 7 =
612 ÷ 4 =
960 ÷ 8 =
56 × 7 =

5 threes plus 3 threes equals 8 threes — 29

Date: _____
Day of Week: _____

STEP 1 — Warm-up (1 min)

Answer these.

34 + 21 = ☐   27 − 13 = ☐   41 + 28 = ☐   38 − 18 = ☐

16 + 34 = ☐   30 + 58 = ☐   56 − 39 = ☐   75 − 38 = ☐

56 + 44 = ☐   88 − 59 = ☐   63 − 36 = ☐   37 + 47 = ☐

STEP 2 — Rapid calculation (2.5 min)

1. Fill in the missing numbers.

   5 × 2 + 3 × 2 = ☐ × 2 = ☐   3 × 3 + 6 × 3 = ☐ × 3 = ☐

   6 × 4 + 4 × 4 = ☐ × 4 = ☐   4 × 5 + 2 × 5 = ☐ × 5 = ☐

   6 × 6 + 3 × 6 = ☐ × 6 = ☐   4 × 7 + 2 × 7 = ☐ × 7 = ☐

2. Answer these.

   3 × 8 + 4 × 8 = ☐   6 × 9 + 2 × 9 = ☐

   5 × 7 + 2 × 7 = ☐   5 × 2 + 6 × 2 = ☐

   7 × 5 + 5 × 5 = ☐   4 × 4 + 4 × 8 = ☐

STEP 3 — Challenge (1.5 min)

Fill in the missing numbers.

19 × 4 = ☐ × ☐ + ☐ × ☐ = ☐ + ☐ = ☐

12 × 8 = ☐ × ☐ + ☐ × ☐ = ☐ + ☐ = ☐

13 × 4 = ☐ × ☐ + ☐ × ☐ = ☐ + ☐ = ☐

15 × 6 = ☐ × ☐ + ☐ × ☐ = ☐ + ☐ = ☐

Time spent: ____ min ____ sec. Total: ____ out of 28

30 — 5 threes minus 3 threes equals 2 threes (1)

Date: _____
Day of Week: _____

STEP 1 — Warm-up (1 min)

Answer these.

52 + 25 = ☐   43 − 33 = ☐   26 + 35 = ☐   45 − 28 = ☐

24 + 38 = ☐   28 + 40 = ☐   42 − 19 = ☐   63 − 45 = ☐

38 + 27 = ☐   63 − 28 = ☐   72 − 27 = ☐   46 + 37 = ☐

STEP 2 — Rapid calculation (2.5 min)

Fill in the missing numbers.

5 × 2 − 3 × 2 = ☐ × 2 = ☐   6 × 3 − 2 × 3 = ☐ × 3 = ☐

5 × 4 − 2 × 4 = ☐ × 4 = ☐   6 × 5 − 4 × 5 = ☐ × 5 = ☐

7 × 6 − 1 × 6 = ☐ × ☐ = ☐   8 × 7 − 4 × 7 = ☐ × ☐ = ☐

8 × 5 − 3 × 5 = ☐ × ☐ = ☐   7 × 9 − 2 × 9 = ☐ × ☐ = ☐

6 × 7 − 2 × 7 = ☐ × ☐ = ☐   18 × 8 − 6 × 8 = ☐ × ☐ = ☐

17 × 3 − 6 × 3 = ☐ × ☐ = ☐   20 × 9 − 8 × 9 = ☐ × ☐ = ☐

STEP 3 — Challenge (1.5 min)

Fill in the missing numbers.

8 × 4 + 3 × 4 − 5 × 4 = ☐ × 4 = ☐

9 × 5 − 3 × 5 + 6 × 5 = ☐ × ☐ = ☐

20 × 7 − 7 − 8 × 7 = ☐ × ☐ = ☐

15 × 9 − 9 × 6 + 9 = ☐ × ☐ = ☐

Mind Gym

※ + ◉ = 24

※ − ◉ = 6

※ = ☐

◉ = ☐

Time spent: ___ min ___ sec. Total: ___ out of 28

5 threes minus 3 threes equals 2 threes (2) — 31

Date: _____
Day of Week: _____

STEP 1 Warm-up (1 min)

Answer these.

38 + 12 = ☐   54 − 26 = ☐   9 × 5 = ☐   4 × 8 = ☐

24 ÷ 3 = ☐   42 ÷ 7 = ☐   65 + 26 = ☐   57 − 29 = ☐

32 + 17 = ☐   63 ÷ 9 = ☐   72 ÷ 8 = ☐   0 ÷ 5 = ☐

STEP 2 Rapid calculation (2.5 min)

Fill in the missing numbers.

7 × 9 − 3 × 9 = ☐ × ☐ = ☐   6 × 3 − 4 × 3 = ☐ × ☐ = ☐

5 × 8 − 2 × 8 = ☐ × ☐ = ☐   8 × 7 − 5 × 7 = ☐ × ☐ = ☐

6 × 4 − 3 × 4 = ☐ × ☐ = ☐   9 × 3 − 6 × 3 = ☐ × ☐ = ☐

10 × 5 − 3 × 5 = ☐ × ☐ = ☐   9 × 9 − 4 × 9 = ☐ × ☐ = ☐

8 × 4 − 7 × 4 = ☐ × ☐ = ☐   9 × 10 − 7 × 10 = ☐ × ☐ = ☐

13 × 9 − 9 = ☐ × ☐ = ☐   20 × 7 − 7 × 8 = ☐ × ☐ = ☐

STEP 3 Challenge (1.5 min)

Fill in the missing numbers.

10 × 9 − 2 × 9 − 3 × 3 = ☐ × ☐ = ☐

11 × 8 + 8 − 2 × 4 = ☐ × ☐ = ☐

10 × 6 − 6 − 3 × 2 × 3 = ☐ × ☐ = ☐

3 × 2 × 5 + 2 × 5 + 2 × 5 × 2 = ☐ × ☐ = ☐

Time spent: _____ min _____ sec. Total: _____ out of 28

32 Practice (1) – Multiples

Date: _____
Day of Week: _____

STEP 1 — Warm-up (1 min)

Answer these.

172 − 72 = ☐   2 × 85 = ☐   270 + 55 = ☐   387 ÷ 3 = ☐

275 × 5 = ☐   510 ÷ 3 = ☐   4 × 29 = ☐   123 × 3 = ☐

844 ÷ 4 = ☐   130 × 4 = ☐

STEP 2 — Rapid calculation (2.5 min)

Answer these.

45 × 5 = ☐   223 × 3 = ☐   60 × 13 = ☐

41 × 8 = ☐   37 × 8 = ☐   74 × 3 = ☐

340 × 4 = ☐   39 × 4 = ☐   54 × 12 = ☐

46 × 7 = ☐   62 × 9 = ☐   78 × 3 = ☐

28 × 5 = ☐   62 × 8 = ☐   83 × 8 = ☐

STEP 3 — Challenge (1.5 min)

Answer these.

420 × 5 = ☐   940 ÷ 4 = ☐

728 ÷ 8 = ☐   162 × 5 = ☐

13 × 17 = ☐   840 ÷ 6 = ☐

235 ÷ 5 = ☐   56 × 6 = ☐

Time spent: _____ min _____ sec. Total: _____ out of 33

Practice (2) – Multiples   33

Date: _____
Day of Week: _____

STEP 1 — Warm-up (1 min)

Answer these.

98 ÷ 7 = ☐   25 × 5 = ☐   278 + 21 = ☐   675 ÷ 9 = ☐

73 × 5 = ☐   162 × 4 = ☐   99 ÷ 9 = ☐   750 ÷ 3 = ☐

205 ÷ 5 = ☐   46 × 8 = ☐

STEP 2 — Rapid calculation (2.5 min)

Answer these.

32 × 9 = ☐   62 × 6 = ☐   71 × 9 = ☐

24 × 12 = ☐   120 × 4 = ☐   85 × 7 = ☐

44 × 6 = ☐   56 × 6 = ☐   35 × 4 = ☐

92 × 5 = ☐   24 × 9 = ☐   27 × 7 = ☐

32 × 5 = ☐   43 × 11 = ☐   75 × 8 = ☐

STEP 3 — Challenge (1.5 min)

Answer these.

335 ÷ 5 = ☐   812 ÷ 4 = ☐

81 × 9 = ☐   32 × 8 = ☐

16 × 25 = ☐   66 × 9 = ☐

568 ÷ 8 = ☐   336 ÷ 6 = ☐

©HarperCollins*Publishers* 2019   Time spent: ___ min ___ sec. Total: ___ out of 33

34 Multiplying whole tens by a two-digit number (1)

Date: _____
Day of Week: _____

STEP 1 Warm-up

Answer these.

30 × 30 = ☐   70 × 60 = ☐   50 × 40 = ☐   20 × 30 = ☐

60 × 80 = ☐   60 × 90 = ☐   90 × 30 = ☐   20 × 70 = ☐

40 × 80 = ☐   50 × 50 = ☐

STEP 2 Rapid calculation

Answer these.

16 × 30 = ☐   47 × 20 = ☐   12 × 40 = ☐

19 × 30 = ☐   25 × 40 = ☐   13 × 20 = ☐

24 × 40 = ☐   22 × 50 = ☐   80 × 12 = ☐

70 × 13 = ☐   30 × 25 = ☐   90 × 11 = ☐

24 × 30 = ☐   30 × 17 = ☐   20 × 48 = ☐

STEP 3 Challenge

Answer these.

20 × 40 = ☐   25 × 20 = ☐   15 × 30 = ☐

13 × 40 = ☐   16 × 90 = ☐   16 × 60 = ☐

17 × 40 = ☐   16 × 80 = ☐   15 × 70 = ☐

Time spent: _____ min _____ sec. Total: _____ out of 34

Multiplying whole tens by a two-digit number (2)

35

Date: _____
Day of Week: _____

STEP 1 — Warm-up (1 min)

Start the timer

Answer these.

3 × 11 = ☐   4 × 12 = ☐   5 × 14 = ☐   6 × 18 = ☐

7 × 12 = ☐   9 × 18 = ☐   6 × 13 = ☐   5 × 19 = ☐

14 × 4 = ☐   13 × 9 = ☐

STEP 2 — Rapid calculation (2.5 min)

Start the timer

Answer these.

30 × 11 = ☐   40 × 12 = ☐   50 × 14 = ☐

60 × 18 = ☐   70 × 12 = ☐   90 × 18 = ☐

60 × 13 = ☐   16 × 20 = ☐   47 × 20 = ☐

12 × 30 = ☐   38 × 20 = ☐   20 × 41 = ☐

19 × 30 = ☐   35 × 20 = ☐   32 × 20 = ☐

STEP 3 — Challenge (1.5 min)

Start the timer

Answer these.

45 × 90 = ☐   47 × 40 = ☐

19 × 20 = ☐   43 × 20 = ☐

43 × 40 = ☐   20 × 55 = ☐

40 × 35 = ☐   25 × 90 = ☐

Mind Gym

1. Make the largest possible product using the digits 2, 4, 6, 8 once only.

   ☐ ☐ ☐ × ☐

2. Make the smallest possible product using the digits 2, 4, 6, 8 once only.

   ☐ ☐ ☐ × ☐

Time spent: _____ min _____ sec. Total: _____ out of 33

36 Multiplying whole tens by a two-digit number (3)

Date: _____

Day of Week: _____

STEP 1 Warm-up

Answer these.

30 × 60 =   70 × 80 =   50 × 50 =   90 × 20 =

40 × 40 =   60 × 50 =   70 × 30 =   20 × 80 =

30 × 50 =   60 × 80 =

STEP 2 Rapid calculation

Answer these.

22 × 40 =   30 × 15 =   12 × 80 =   59 × 20 =

38 × 30 =   34 × 30 =   16 × 60 =   17 × 70 =

16 × 30 =   28 × 40 =   30 × 35 =   20 × 26 =

50 × 15 =   26 × 50 =   70 × 19 =   13 × 30 =

14 × 20 =   15 × 20 =   18 × 40 =   21 × 40 =

STEP 3 Challenge

Answer these.

23 × 40 =   65 × 20 =   45 × 20 =

11 × 90 =   60 × 12 =   23 × 30 =

125 × 20 =   15 × 50 =   15 × 60 =

38 × 20 =   33 × 30 =   41 × 40 =

Time spent: _____ min _____ sec. Total: _____ out of 42

Multiplying a two-digit number by a two-digit number (1) 37

STEP 1 — Warm-up

Answer these.

1. 12 × 2 = ☐
   12 × 20 = ☐

2. 24 × 3 = ☐
   24 × 30 = ☐

3. 13 × 2 = ☐
   13 × 20 = ☐

4. 32 × 5 = ☐
   32 × 50 = ☐

5. 29 × 7 = ☐
   29 × 70 = ☐

STEP 2 — Rapid calculation

Answer these. The first one has been done for you.

1. 25 × 12
   = 25 × **10** + 25 × **2**
   = **250** + **50**
   = **300**

2. 49 × 23
   = 49 × ☐ + 49 × ☐
   = ☐ + ☐
   = ☐

3. 35 × 46
   = ☐ × ☐ + ☐ × ☐
   = ☐ + ☐
   = ☐

4. 21 × 18
   = ☐ × ☐ + ☐ × ☐
   = ☐ + ☐
   = ☐

5. 29 × 52
   = ☐ × ☐ + ☐ × ☐
   = ☐ + ☐
   = ☐

6. 71 × 34
   = ☐ × ☐ + ☐ × ☐
   = ☐ + ☐
   = ☐

STEP 3 — Challenge

Answer these.

15 × 11 = ☐   25 × 11 = ☐   11 × 38 = ☐   11 × 81 = ☐

11 × 34 = ☐   14 × 11 = ☐   11 × 63 = ☐   11 × 92 = ☐

Time spent: ____ min ____ sec. Total: ____ out of 23

38 Multiplying a two-digit number by a two-digit number (2)

Date: _____
Day of Week: _____

STEP 1 (1 min) Warm-up

Answer these.

4 × 19 = ☐   76 × 4 = ☐   35 × 5 = ☐   11 × 7 = ☐

39 × 4 = ☐   78 × 6 = ☐   27 × 5 = ☐   37 × 3 = ☐

STEP 2 (2.5 min) Rapid calculation

Fill in the missing numbers. The first one has been done for you.

24 × 36 = 24 × **30** + 24 × **6** = **720** + **144** = **864**

37 × 13 = 37 × ☐ + 37 × ☐ = ☐ + ☐ = ☐

16 × 25 = 16 × ☐ + 16 × ☐ = ☐ + ☐ = ☐

41 × 18 = 41 × ☐ + 41 × ☐ = ☐ + ☐ = ☐

38 × 12 = ☐ × ☐ + ☐ × ☐ = ☐ + ☐ = ☐

84 × 31 = ☐ × ☐ + ☐ × ☐ = ☐ + ☐ = ☐

74 × 23 = ☐ × ☐ + ☐ × ☐ = ☐ + ☐ = ☐

52 × 14 = ☐ × ☐ + ☐ × ☐ = ☐ + ☐ = ☐

46 × 27 = ☐ × ☐ + ☐ × ☐ = ☐ + ☐ = ☐

68 × 26 = ☐ × ☐ + ☐ × ☐ = ☐ + ☐ = ☐

Mind Gym

Move one matchstick to make the number sentence true.

STEP 3 (1.5 min) Challenge

Answer these.

11 × 74 = ☐   78 × 11 = ☐   46 × 11 = ☐   63 × 11 = ☐

54 × 11 = ☐   49 × 11 = ☐   69 × 11 = ☐   86 × 11 = ☐

Time spent: _____ min _____ sec. Total: _____ out of 25

Multiplying a two-digit number by a two-digit number (3)

Date: _____
Day of Week: _____

STEP 1 — Warm-up (1 min)

Answer these.

76 × 20 = ☐   54 × 80 = ☐   18 × 40 = ☐   49 × 20 = ☐

57 × 30 = ☐   38 × 40 = ☐   73 × 30 = ☐   26 × 50 = ☐

STEP 2 — Rapid calculation (2.5 min)

Fill in the missing numbers.

31 × 22 = ☐ × ☐ + ☐ × ☐ = ☐ + ☐ = ☐

36 × 14 = ☐ × ☐ + ☐ × ☐ = ☐ + ☐ = ☐

18 × 25 = ☐ × ☐ + ☐ × ☐ = ☐ + ☐ = ☐

43 × 17 = ☐ × ☐ + ☐ × ☐ = ☐ + ☐ = ☐

38 × 16 = ☐ × ☐ + ☐ × ☐ = ☐ + ☐ = ☐

77 × 41 = ☐ × ☐ + ☐ × ☐ = ☐ + ☐ = ☐

87 × 31 = ☐ × ☐ + ☐ × ☐ = ☐ + ☐ = ☐

64 × 25 = ☐ × ☐ + ☐ × ☐ = ☐ + ☐ = ☐

53 × 18 = ☐ × ☐ + ☐ × ☐ = ☐ + ☐ = ☐

94 × 31 = ☐ × ☐ + ☐ × ☐ = ☐ + ☐ = ☐

STEP 3 — Challenge (1.5 min)

TIP: When a two-digit number is multiplied by 15, multiply it by 10 and then add half the result.
For example: 18 × 15 = 180 + 180 ÷ 2
            = 180 + 90 = 270

Answer these.

15 × 70 = ☐   15 × 42 = ☐   15 × 38 = ☐   46 × 15 = ☐

24 × 15 = ☐   72 × 15 = ☐   15 × 66 = ☐   15 × 94 = ☐

Time spent: _____ min _____ sec. Total: _____ out of 26

40 Multiplication and division of two-digit numbers

Date: _____
Day of Week: _____

STEP 1 — Warm-up

Answer these.

160 ÷ 40 = ☐   420 ÷ 70 = ☐   960 ÷ 30 = ☐

34 × 50 = ☐   66 × 50 = ☐   25 × 50 = ☐

84 × 80 = ☐   13 × 70 = ☐   210 ÷ 70 = ☐

STEP 2 — Rapid calculation

Answer these.

700 ÷ 20 = ☐   570 ÷ 30 = ☐   90 × 46 = ☐

12 × 90 = ☐   390 ÷ 30 = ☐   87 × 30 = ☐

45 × 50 = ☐   960 ÷ 80 = ☐   760 ÷ 40 = ☐

900 ÷ 50 = ☐   440 ÷ 20 = ☐   22 × 60 = ☐

630 ÷ 90 = ☐   540 ÷ 30 = ☐   56 × 70 = ☐

STEP 3 — Challenge

Answer these.

700 ÷ 70 = ☐   570 ÷ 30 = ☐

520 ÷ 4 = ☐   480 ÷ 8 = ☐

420 × 9 = ☐   30 × 50 = ☐

900 ÷ 30 = ☐   720 ÷ 30 = ☐

Time spent: _____ min _____ sec. Total: _____ out of 32

Multiplying a three-digit number by a one-digit number (1)

Date: _____
Day of Week: _____

STEP 1 Warm-up

Answer these.

1. 15 × 3 = ☐
   150 × 3 = ☐

2. 35 × 4 = ☐
   350 × 4 = ☐

3. 16 × 8 = ☐
   160 × 8 = ☐

4. 36 × 6 = ☐
   360 × 6 = ☐

5. 27 × 4 = ☐
   270 × 4 = ☐

STEP 2 Rapid calculation

Answer these.

320 × 3 = ☐   270 × 4 = ☐   190 × 6 = ☐

430 × 3 = ☐   540 × 4 = ☐   630 × 5 = ☐

160 × 8 = ☐   440 × 5 = ☐   370 × 3 = ☐

640 × 5 = ☐   270 × 7 = ☐   260 × 6 = ☐

140 × 8 = ☐   210 × 9 = ☐   510 × 8 = ☐

STEP 3 Challenge

Answer these.

330 × 7 = ☐   440 × 5 = ☐

290 × 3 = ☐   560 × 4 = ☐

670 × 5 = ☐   230 × 9 = ☐

860 × 8 = ☐   480 × 8 = ☐

Time spent: ____ min ____ sec. Total: ____ out of 33

42 Multiplying a three-digit number by a one-digit number (2)

Date: _____
Day of Week: _____

STEP 1 Warm-up

Answer these.

1. 32 × 3 = ☐   2. 28 × 4 = ☐   3. 9 × 16 = ☐

   320 × 3 = ☐   280 × 4 = ☐   9 × 160 = ☐

4. 24 × 6 = ☐   5. 5 × 38 = ☐

   240 × 6 = ☐   5 × 380 = ☐

STEP 2 Rapid calculation

Answer these.

270 × 3 = ☐   670 × 4 = ☐   560 × 5 = ☐

450 × 7 = ☐   720 × 4 = ☐   840 × 4 = ☐

590 × 3 = ☐   950 × 3 = ☐   370 × 5 = ☐

390 × 8 = ☐   640 × 7 = ☐   360 × 6 = ☐

260 × 8 = ☐   330 × 9 = ☐   610 × 7 = ☐

STEP 3 Challenge

Answer these.

540 × 4 = ☐   560 × 3 = ☐

650 × 5 = ☐   460 × 6 = ☐

790 × 4 = ☐   580 × 8 = ☐

380 × 7 = ☐   270 × 9 = ☐

Time spent: _____ min _____ sec.   Total: _____ out of 33

Dividing three-digit numbers by tens (1)

43

Date: _____
Day of Week: _____

STEP 1 — Warm-up (1 min)

Answer these.

400 ÷ 2 = ☐   200 ÷ 5 = ☐   600 ÷ 5 = ☐   300 ÷ 6 = ☐

810 ÷ 9 = ☐   420 ÷ 7 = ☐   320 ÷ 4 = ☐   640 ÷ 8 = ☐

540 ÷ 9 = ☐   120 ÷ 4 = ☐

STEP 2 — Rapid calculation (2.5 min)

Answer these.

400 ÷ 20 = ☐   300 ÷ 60 = ☐   500 ÷ 50 = ☐

400 ÷ 80 = ☐   600 ÷ 10 = ☐   120 ÷ 20 = ☐

240 ÷ 30 = ☐   480 ÷ 60 = ☐   560 ÷ 80 = ☐

350 ÷ 70 = ☐   810 ÷ 90 = ☐   360 ÷ 60 = ☐

540 ÷ 60 = ☐   390 ÷ 30 = ☐   280 ÷ 70 = ☐

STEP 3 — Challenge (1.5 min)

Answer these.

720 ÷ 80 = ☐   180 ÷ 20 = ☐

420 ÷ 60 = ☐   280 ÷ 20 = ☐

640 ÷ 80 = ☐   490 ÷ 70 = ☐

300 ÷ 50 = ☐   840 ÷ 20 = ☐

Time spent: _____ min _____ sec. Total: _____ out of 33

44 Dividing three-digit numbers by tens (2)

Date: _____
Day of Week: _____

STEP 1 Warm-up

Answer these.

240 ÷ 60 = ☐   480 ÷ 80 = ☐   360 ÷ 60 = ☐

360 ÷ 40 = ☐   560 ÷ 70 = ☐   240 ÷ 40 = ☐

300 ÷ 60 = ☐   250 ÷ 50 = ☐   360 ÷ 90 = ☐

STEP 2 Rapid calculation

Answer these.

630 ÷ 90 = ☐   540 ÷ 90 = ☐   630 ÷ 70 = ☐

320 ÷ 80 = ☐   260 ÷ 20 = ☐   640 ÷ 80 = ☐

600 ÷ 60 = ☐   450 ÷ 50 = ☐   400 ÷ 50 = ☐

540 ÷ 30 = ☐   180 ÷ 20 = ☐   480 ÷ 60 = ☐

480 ÷ 40 = ☐   560 ÷ 80 = ☐   500 ÷ 20 = ☐

STEP 3 Challenge

Answer these.

750 ÷ 50 = ☐   500 ÷ 50 = ☐

700 ÷ 20 = ☐   800 ÷ 20 = ☐

930 ÷ 30 = ☐   240 ÷ 60 = ☐

990 ÷ 30 = ☐   900 ÷ 50 = ☐

Time spent: _____ min _____ sec. Total: _____ out of 32

Dividing a three-digit number by a one-digit number (1)

Date: _____
Day of Week: _____

STEP 1 Warm-up

Answer these.

624 ÷ 6 = ☐   270 ÷ 6 = ☐

618 ÷ 6 = ☐   345 ÷ 5 = ☐

936 ÷ 3 = ☐   161 ÷ 7 = ☐

256 ÷ 8 = ☐   768 ÷ 4 = ☐

STEP 2 Rapid calculation

Answer these.

369 ÷ 3 = ☐   468 ÷ 2 = ☐   752 ÷ 8 = ☐

864 ÷ 6 = ☐   645 ÷ 5 = ☐   954 ÷ 6 = ☐

405 ÷ 5 = ☐   133 ÷ 7 = ☐   963 ÷ 9 = ☐

700 ÷ 5 = ☐   330 ÷ 6 = ☐   378 ÷ 7 = ☐

356 ÷ 4 = ☐   207 ÷ 3 = ☐   504 ÷ 9 = ☐

STEP 3 Challenge

Answer these.

104 ÷ 4 = ☐   324 ÷ 3 = ☐   150 ÷ 6 = ☐

256 ÷ 8 = ☐   666 ÷ 6 = ☐   465 ÷ 5 = ☐

756 ÷ 9 = ☐   567 ÷ 7 = ☐   900 ÷ 9 = ☐

Time spent: _____ min _____ sec. Total: _____ out of 32

46 Dividing a three-digit number by a one-digit number (2)

Date: _____
Day of Week: _____

STEP 1 (1 min) Warm-up

Answer these.

400 ÷ 4 = ☐   880 ÷ 8 = ☐   990 ÷ 9 = ☐   604 ÷ 2 = ☐

873 ÷ 3 = ☐   189 ÷ 9 = ☐   315 ÷ 5 = ☐   255 ÷ 3 = ☐

STEP 2 (2.5 min) Rapid calculation

Answer these.

500 ÷ 2 = ☐   260 ÷ 5 = ☐   365 ÷ 5 = ☐

765 ÷ 3 = ☐   630 ÷ 3 = ☐   450 ÷ 5 = ☐

174 ÷ 3 = ☐   840 ÷ 7 = ☐   903 ÷ 3 = ☐

624 ÷ 4 = ☐   666 ÷ 3 = ☐   990 ÷ 6 = ☐

777 ÷ 7 = ☐   912 ÷ 6 = ☐   952 ÷ 2 = ☐

STEP 3 (1.5 min) Challenge

Fill in the missing numbers.

☐ ÷ 7 = 21   350 ÷ ☐ = 5

☐ ÷ 9 = 52   630 ÷ ☐ = 3

☐ ÷ 3 = 99   ☐ ÷ 4 = 130 + 26

749 ÷ ☐ = 63 − 56   320 ÷ ☐ = 8 × 5

Time spent: ____ min ____ sec. Total: ____ out of 31

Column multiplication (1) 47

Date: _____
Day of Week: _____

STEP 1 (1 min) Warm-up

Answer these.

4 × 53 = ☐   37 × 6 = ☐   48 × 4 = ☐   8 × 22 = ☐

61 × 7 = ☐   5 × 76 = ☐   39 × 6 = ☐   7 × 64 = ☐

STEP 2 (2.5 min) Rapid calculation

Estimate first, then use the expanded written method to work out the answers. Remember to use your estimate to check your answer. The first one has been done for you.

1. 76 × 3 = **228**
   Estimate: **240**

   ```
         7 6
       ×   3
       -----
         1 8
       2 1 0
       -----
       2 2 8
   ```

2. 87 × 6 =
   Estimate:
   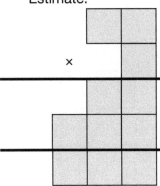

3. 94 × 8 =
   Estimate:
   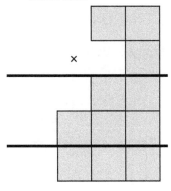

4. 317 × 4 =
   Estimate:
   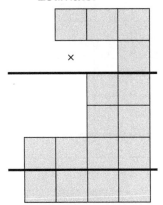

5. 456 × 3 =
   Estimate:
   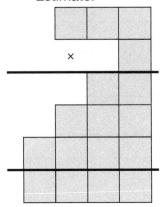

6. 748 × 5 =
   Estimate:
   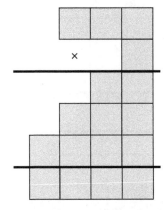

STEP 3 (1.5 min) Challenge

Use the expanded written method to answer these. Use a separate sheet of paper for your working.

28 × 6 = ☐   7 × 37 = ☐   53 × 8 = ☐

4 × 426 = ☐   578 × 6 = ☐   8 × 664 = ☐

48 Column multiplication (2)

Date: _____
Day of Week: _____

STEP 1 — Warm-up

Answer these.

3 × 78 = ☐   84 × 5 = ☐   66 × 7 = ☐   6 × 92 = ☐

87 × 9 = ☐   8 × 68 = ☐   93 × 7 = ☐   6 × 83 = ☐

STEP 2 — Rapid calculation

Estimate first, then use short multiplication to work out the answers. Remember to use your estimate to check your answer. The first one has been done for you.

1. 67 × 5 = **335**
   Estimate: **350**
   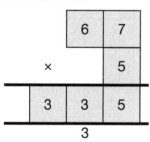

2. 56 × 9 =
   Estimate:
   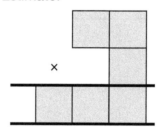

3. 93 × 8 =
   Estimate:
   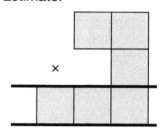

4. 247 × 6 =
   Estimate:
   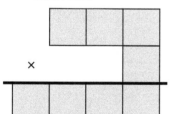

5. 486 × 7 =
   Estimate:
   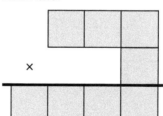

6. 728 × 8 =
   Estimate:
   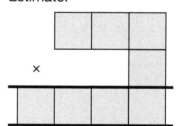

STEP 3 — Challenge

Use short multiplication to answer these. Use a separate sheet of paper for your working.

59 × 6 = ☐   8 × 73 = ☐   88 × 9 = ☐

4 × 358 = ☐   678 × 6 = ☐   8 × 894 = ☐

Time spent: _____ min _____ sec. Total: _____ out of 19

Relationship between multiplication and division (1)

Date: _____
Day of Week: _____

STEP 1 Warm-up

Answer these.

1. 60 × 8 = ☐
   480 ÷ 60 = ☐
   480 ÷ 8 = ☐

2. 50 × 6 = ☐
   300 ÷ 50 = ☐
   300 ÷ 6 = ☐

3. 14 × 15 = ☐
   210 ÷ 14 = ☐
   210 ÷ 15 = ☐

4. 16 × 60 = ☐
   960 ÷ 60 = ☐
   960 ÷ 16 = ☐

STEP 2 Rapid calculation

Remember the inverse relationship between multiplication and division.

Fill in the missing numbers.

☐ ÷ 12 = 8   95 ÷ ☐ = 5   360 ÷ ☐ = 9
100 × ☐ = 1000   ☐ × 24 = 120   ☐ × 4 = 720
☐ ÷ 86 = 30   1000 ÷ ☐ = 8   ☐ × 5 = 80
165 ÷ ☐ = 11   ☐ × 12 = 600   ☐ × 40 = 960
720 ÷ ☐ = 30   50 × ☐ = 900   450 ÷ ☐ = 30

STEP 3 Challenge

Fill in the missing numbers.

☐ × 12 = 200 + 40   ☐ × 4 = 120 × 2
☐ × 8 = 360 ÷ 9   ☐ ÷ 25 = 5 × 5
☐ × 14 = 6 × 7 × 5   36 × ☐ = 72 × 20
570 ÷ ☐ = 5 × 6   ☐ ÷ 28 = 72 ÷ 9

50 Relationship between multiplication and division (2)

STEP 1 Warm-up

Answer these.

13 × 80 = ☐   154 ÷ 7 = ☐   208 ÷ 8 = ☐   120 × 5 = ☐

600 ÷ 5 = ☐   642 ÷ 6 = ☐   336 ÷ 6 = ☐   16 × 15 = ☐

520 ÷ 8 = ☐   32 × 60 = ☐

STEP 2 Rapid calculation

Fill in the missing numbers.

☐ ÷ 32 = 11   175 ÷ ☐ = 5   108 ÷ ☐ = 9

☐ ÷ 20 = 25   520 ÷ ☐ = 40   ☐ × 24 = 120

☐ × 4 = 144   ☐ ÷ 24 = 15   2000 ÷ ☐ = 8

☐ ÷ 23 = 11   720 ÷ ☐ = 30   28 × ☐ = 140

☐ × 5 = 600   ☐ × 80 = 3200   45 × ☐ = 900

STEP 3 Challenge

Fill in the missing numbers.

☐ ÷ 26 = 50   ☐ × 18 = 320 + 40

14 × ☐ = 12 × 28   ☐ × 12 = 540 ÷ 9

☐ ÷ 25 = 24 × 5   ☐ × 18 = 54 × 7

960 ÷ ☐ = 5 × 24   ☐ ÷ 28 = 112 − 98

Fractions (1)

STEP 1 — Warm-up (1 min)

Write a fraction to describe the shaded part.

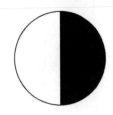  

STEP 2 — Rapid calculation (2.5 min)

Fill in the missing numbers.

$\frac{1}{2}$ of 12 ☆s is ☐ ☆s. $\frac{1}{5}$ of 25 ☆s is ☐ ☆s. $\frac{1}{4}$ of 20 ◎s is ☐ ◎s.

$\frac{1}{2}$ of 10 ☐s is ☐ ☐s. $\frac{1}{7}$ of 14 △s is ☐ △s. $\frac{1}{4}$ of 16 ◇s is ☐ ◇s.

$\frac{1}{6}$ of 24 ◎s is ☐ ◎s. $\frac{1}{9}$ of 36 ■s is ☐ ■s. $\frac{1}{5}$ of 40 △s is ☐ △s.

STEP 3 — Challenge (1.5 min)

Write **T** if the fraction matches the darker-shaded area of the shape or set, and **F** if it does not.

1. $\frac{1}{4}$ ☐
2. $\frac{1}{3}$ ☐
3. $\frac{1}{2}$ ☐
4. $\frac{1}{4}$ ☐

5. $\frac{1}{2}$ ☐
6. $\frac{1}{4}$ ☐
7. $\frac{1}{3}$ ☐
8. $\frac{1}{3}$ ☐

Time spent: _____ min _____ sec.   Total: _____ out of 22

52 Fractions (2)

Date: _____
Day of Week: _____

STEP 1 (1 min) Warm-up

Write fractions to describe the shaded parts.

STEP 2 (2.5 min) Rapid calculation

Fill in the missing numbers.

$\frac{2}{3}$ of 6 🍎 s is ☐ 🍎 s. $\frac{1}{4}$ of 12 🍌 s is ☐ 🍌 s.

$\frac{2}{5}$ of 25 🍓 s is ☐ 🍓 s. $\frac{3}{8}$ of 32 🍍 s is ☐ 🍍 s.

$\frac{3}{4}$ of 24 🍐 s is ☐ 🍐 s. $\frac{3}{7}$ of 28 🍒 s is ☐ 🍒 s.

$\frac{2}{9}$ of 27 bunches of 🍇 s is ☐ bunches of 🍇 s. $\frac{5}{8}$ of 40 🍉 s is ☐ 🍉 s.

$\frac{1}{6}$ of 18 🍑 s is ☐ 🍑 s. $\frac{3}{5}$ of 45 ⚽ s is ☐ ⚽ s.

STEP 3 (1.5 min) Challenge

Write fractions to describe the darker-shaded parts.

Time spent: _____ min _____ sec. Total: _____ out of 25

Addition of fractions

STEP 1 Warm-up (1 min)

Fill in the missing numbers.

$\dfrac{3}{7} + \dfrac{2}{7} = \dfrac{3 + \square}{7} = \dfrac{\square}{7}$

$\dfrac{1}{5} + \dfrac{3}{5} = \dfrac{\square + \square}{\square} = \dfrac{\square}{\square}$

$\dfrac{5}{10} + \dfrac{4}{10} = \dfrac{\square + \square}{\square} = \dfrac{\square}{\square}$

$\dfrac{13}{25} + \dfrac{11}{25} = \dfrac{\square}{\square} = \dfrac{\square}{\square}$

$\dfrac{2}{8} + \dfrac{5}{8} = \dfrac{\square}{\square} = \dfrac{\square}{\square}$

$\dfrac{28}{123} + \dfrac{49}{123} = \dfrac{\square}{\square} = \dfrac{\square}{\square}$

STEP 2 Rapid calculation (2.5 min)

TIP *When adding or subtracting fractions with the same denominator, simply add or subtract the numerators and leave the denominator unchanged.*

Answer these.

$\dfrac{5}{9} + \dfrac{2}{9} =$ 　　 $\dfrac{5}{58} + \dfrac{34}{58} =$ 　　 $\dfrac{11}{20} + \dfrac{4}{20} =$ 　　 $\dfrac{3}{11} + \dfrac{5}{11} =$

$\dfrac{5}{12} + \dfrac{4}{12} =$ 　　 $\dfrac{14}{19} + \dfrac{2}{19} =$ 　　 $\dfrac{4}{15} + \dfrac{6}{15} =$ 　　 $\dfrac{8}{17} + \dfrac{7}{17} =$

$\dfrac{13}{47} + \dfrac{24}{47} =$ 　　 $\dfrac{3}{8} + \dfrac{4}{8} =$ 　　 $\dfrac{15}{39} + \dfrac{22}{39} =$ 　　 $\dfrac{11}{20} + \dfrac{2}{20} =$

$\dfrac{53}{108} + \dfrac{24}{108} =$ 　　 $\dfrac{28}{209} + \dfrac{12}{209} =$ 　　 $\dfrac{28}{107} + \dfrac{57}{107} =$

STEP 3 Challenge (1.5 min)

Fill in the missing fractions.

$\dfrac{2}{10} + \dfrac{3}{10} + \dfrac{4}{10} = \square$ 　　 $\dfrac{5}{17} + \dfrac{7}{17} + \dfrac{3}{17} = \square$

$\dfrac{25}{137} + \dfrac{36}{137} + \dfrac{40}{137} = \square$ 　　 $\dfrac{12}{29} + \dfrac{3}{29} + \dfrac{4}{29} = \square$

$\dfrac{5}{18} + \square = \dfrac{1}{2}$ 　　 $\dfrac{5}{24} + \square = \dfrac{1}{4}$

$\square + \dfrac{17}{225} = \dfrac{1}{5}$ 　　 $\square + \dfrac{7}{108} = \dfrac{1}{6}$

Time spent: _____ min _____ sec. Total: _____ out of 29

54 Subtraction of fractions

Date: _____
Day of Week: _____

STEP 1 (1 min) Warm-up

Fill in the missing numbers.

$\dfrac{3}{7} - \dfrac{2}{7} = \dfrac{3 - \square}{\square \, 7} = \dfrac{\square}{\square \, 7}$ $\dfrac{3}{5} - \dfrac{1}{5} = \dfrac{\square - \square}{\square} = \dfrac{\square}{\square}$ $\dfrac{5}{10} - \dfrac{4}{10} = \dfrac{\square - \square}{\square} = \dfrac{\square}{\square}$

$\dfrac{5}{8} - \dfrac{2}{8} = \dfrac{\square}{\square} = \dfrac{\square}{\square}$ $\dfrac{13}{19} - \dfrac{8}{19} = \dfrac{\square}{\square} = \dfrac{\square}{\square}$ $\dfrac{11}{14} - \dfrac{3}{14} = \dfrac{\square}{\square} = \dfrac{\square}{\square}$

$\dfrac{15}{18} - \dfrac{7}{18} = \dfrac{\square}{\square} = \dfrac{\square}{\square}$ $\dfrac{28}{29} - \dfrac{12}{29} = \dfrac{\square}{\square} = \dfrac{\square}{\square}$ $\dfrac{13}{16} - \dfrac{7}{16} = \dfrac{\square}{\square} = \dfrac{\square}{\square}$

STEP 2 (2.5 min) Rapid calculation

Answer these.

$\dfrac{6}{7} - \dfrac{2}{7} = \square$ $\dfrac{9}{10} - \dfrac{4}{10} = \square$ $\dfrac{26}{51} - \dfrac{18}{51} = \square$

$\dfrac{9}{10} - \dfrac{7}{10} = \square$ $\dfrac{39}{47} - \dfrac{22}{47} = \square$ $\dfrac{9}{18} - \dfrac{1}{18} = \square$

$\dfrac{11}{15} - \dfrac{4}{15} = \square$ $\dfrac{8}{11} - \dfrac{3}{11} = \square$ $\dfrac{5}{7} - \dfrac{4}{7} = \square$

$\dfrac{7}{11} - \dfrac{3}{11} = \square$ $\dfrac{15}{16} - \dfrac{12}{16} = \square$ $\dfrac{3}{6} - \dfrac{1}{6} = \square$

$\dfrac{93}{107} - \dfrac{49}{107} = \square$ $\dfrac{29}{38} - \dfrac{14}{38} = \square$ $\dfrac{72}{83} - \dfrac{34}{83} = \square$

STEP 3 (1.5 min) Challenge

Fill in the missing fractions.

$\dfrac{6}{7} - \dfrac{2}{7} - \dfrac{2}{7} = \square$ $\dfrac{29}{38} - \dfrac{12}{38} - \dfrac{5}{38} = \square$ $\dfrac{85}{104} - \dfrac{21}{104} - \dfrac{13}{104} = \square$

$\dfrac{36}{47} - \dfrac{12}{47} - \dfrac{1}{47} = \square$ $\dfrac{49}{100} - \square = \dfrac{2}{5}$ $\square - \dfrac{4}{45} = \dfrac{1}{9}$

$\dfrac{37}{64} - \square = \dfrac{1}{8}$ $\square - \dfrac{5}{72} = \dfrac{1}{2}$

Time spent: _____ min _____ sec. Total: _____ out of 32

Addition and subtraction of fractions

Date: _____
Day of Week: _____

STEP 1 — Warm-up (1 min)

Answer these.

$\dfrac{3}{12} + \dfrac{7}{12} =$ ☐ $\dfrac{11}{18} + \dfrac{7}{18} =$ ☐ $\dfrac{78}{109} - \dfrac{63}{109} =$ ☐

$\dfrac{21}{45} + \dfrac{13}{45} =$ ☐ $\dfrac{34}{71} - \dfrac{18}{71} =$ ☐ $\dfrac{18}{19} - \dfrac{8}{19} =$ ☐

$\dfrac{22}{24} - \dfrac{15}{24} =$ ☐ $\dfrac{12}{37} + \dfrac{17}{37} =$ ☐ $\dfrac{79}{96} - \dfrac{28}{96} =$ ☐

STEP 2 — Rapid calculation (2.5 min)

TIP: If the numerator and the denominator of a fraction are the same, it can be simplified to 1. Therefore, 1 can be changed to any fraction with the same numerator and denominator (except 0) to help with calculations.

Answer these.

$\dfrac{5}{19} + \dfrac{12}{19} =$ ☐ $1 - \dfrac{27}{48} =$ ☐ $\dfrac{25}{28} - \dfrac{18}{28} =$ ☐ $\dfrac{32}{37} - \dfrac{15}{37} =$ ☐

$\dfrac{32}{35} - \dfrac{13}{35} =$ ☐ $\dfrac{13}{24} + \dfrac{7}{24} =$ ☐ $\dfrac{15}{40} + \dfrac{25}{40} =$ ☐ $1 - \dfrac{49}{67} =$ ☐

$\dfrac{8}{27} + \dfrac{19}{27} =$ ☐ $1 - \dfrac{17}{56} =$ ☐ $\dfrac{33}{39} - \dfrac{18}{39} =$ ☐ $1 - \dfrac{24}{27} =$ ☐

$\dfrac{15}{28} + \dfrac{13}{28} =$ ☐ $\dfrac{27}{75} + \dfrac{18}{75} =$ ☐ $\dfrac{19}{30} + \dfrac{6}{30} =$ ☐ $\dfrac{41}{53} + \dfrac{12}{53} =$ ☐

STEP 3 — Challenge (1.5 min)

Answer these.

$\dfrac{4}{15} + \dfrac{7}{15} - \dfrac{3}{15} =$ ☐ $1 - \dfrac{7}{24} - \dfrac{5}{24} =$ ☐ $\dfrac{38}{39} - \dfrac{14}{39} - \dfrac{7}{39} =$ ☐

$\dfrac{5}{42} + \dfrac{17}{42} + \dfrac{13}{42} =$ ☐ $\dfrac{35}{48} - \dfrac{14}{48} + \dfrac{7}{48} =$ ☐ $\dfrac{43}{54} - \dfrac{38}{54} + \dfrac{29}{54} =$ ☐

$\dfrac{4}{15} + \dfrac{11}{15} - \dfrac{23}{47} =$ ☐ $\dfrac{37}{42} + \dfrac{5}{42} - \dfrac{14}{31} =$ ☐

Time spent: ___ min ___ sec. Total: ___ out of 33

56 Practice with fractions (1)

Date: _____
Day of Week: _____

STEP 1 (1 min) Warm-up

Answer these.

$\dfrac{2}{11} + \dfrac{7}{11} = \square$ $\dfrac{5}{13} + \dfrac{8}{13} = \square$ $\dfrac{3}{12} + \dfrac{7}{12} = \square$

$\dfrac{12}{19} + \dfrac{3}{19} = \square$ $\dfrac{12}{31} + \dfrac{17}{31} = \square$ $\dfrac{18}{20} - \dfrac{15}{20} = \square$

$1 - \dfrac{27}{41} = \square$ $\dfrac{25}{47} - \dfrac{25}{47} = \square$ $\dfrac{57}{64} - \dfrac{25}{64} = \square$

STEP 2 (2.5 min) Rapid calculation

Answer these.

$\dfrac{7}{16} + \dfrac{8}{16} = \square$ $1 - \dfrac{29}{40} = \square$ $\dfrac{16}{50} + \dfrac{34}{50} = \square$ $\dfrac{32}{65} + \dfrac{18}{65} = \square$

$\dfrac{17}{26} + \dfrac{8}{26} = \square$ $\dfrac{17}{29} + \dfrac{12}{29} = \square$ $\dfrac{32}{45} - \dfrac{15}{45} = \square$ $\dfrac{25}{34} + \dfrac{9}{34} = \square$

$\dfrac{29}{37} - \dfrac{14}{37} = \square$ $1 - \dfrac{29}{37} = \square$ $1 - \dfrac{17}{30} = \square$ $\dfrac{63}{74} - \dfrac{38}{74} = \square$

$\dfrac{27}{94} + \dfrac{35}{94} = \square$ $\dfrac{23}{29} - \dfrac{15}{29} = \square$ $\dfrac{32}{43} - \dfrac{17}{43} = \square$ $1 - \dfrac{11}{18} = \square$

STEP 3 (1.5 min) Challenge

Fill in the missing fractions.

$1 - \dfrac{7}{16} + \dfrac{8}{16} = \square$ $\dfrac{30}{74} + \dfrac{44}{74} - \dfrac{18}{91} = \square$

$\dfrac{32}{45} - \dfrac{11}{45} - \dfrac{2}{45} = \square$ $\dfrac{24}{93} + \dfrac{23}{93} + \dfrac{14}{93} = \square$

$\dfrac{32}{45} - \square = \dfrac{1}{5}$ $\dfrac{48}{96} - \square = \dfrac{1}{12}$

$\square - \dfrac{3}{84} = \dfrac{1}{21}$ $\square - \dfrac{12}{44} = \dfrac{1}{4}$

Time spent: _____ min _____ sec. Total: _____ out of 33

Practice with fractions (2) 57

STEP 1 — Warm-up

Answer these.

200 ÷ 5 = ☐   99 ÷ 3 = ☐   320 ÷ 8 = ☐   840 ÷ 4 = ☐

660 ÷ 6 = ☐   350 ÷ 7 = ☐   750 ÷ 5 = ☐   540 ÷ 9 = ☐

400 ÷ 8 = ☐   490 ÷ 7 = ☐

STEP 2 — Rapid calculation

Answer these.

$\frac{3}{4}$ of 80 = ☐ $\frac{2}{3}$ of 90 = ☐ $\frac{4}{5}$ of 55 = ☐

$\frac{9}{10}$ of 200 = ☐ $\frac{3}{8}$ of 96 = ☐ $\frac{5}{6}$ of 120 = ☐

$\frac{3}{5}$ of £35 = £☐ $\frac{4}{9}$ of 108 kg = ☐ kg $\frac{3}{4}$ of 88 cm = ☐ cm

$\frac{7}{8}$ of 120 sec = ☐ sec $\frac{5}{12}$ of 144 min = ☐ min $\frac{3}{15}$ of £450 = £☐

STEP 3 — Challenge

Answer these.

$\frac{3}{20}$ of 120 = ☐ $\frac{7}{15}$ of 90 = ☐

$\frac{5}{12}$ of 96 = ☐ $\frac{11}{50}$ of 300 = ☐

$\frac{7}{40}$ of 240 = ☐ $\frac{14}{25}$ of 125 = ☐

$\frac{6}{7}$ of 84 = ☐ $\frac{11}{60}$ of 300 = ☐

58 Decimals in life

Date: _____
Day of Week: _____

STEP 1 — Warm-up (1 min)

Circle the numbers which have decimal places.

0.15   7.23   64   100.04   102   0.3   117   20.1   38   0.04   30.2   33

STEP 2 — Rapid calculation (2.5 min)

Complete these sentences. The first one has been done for you.

0.08 m means [8] centimetres, read as <u>zero point zero eight</u> metres.

1.50 m means [ ] metre [ ] centimetres, read as .... metres.

4.25 m means [ ] metres [ ] centimetres, read as .... metres.

1.20 m means [ ] metre [ ] centimetres, read as .... metres.

2.18 m means [ ] metres [ ] centimetres, read as .... metres.

49.90 m means [ ] metres [ ] centimetres, read as .... metres.

STEP 3 — Challenge (1.5 min)

TIP: $1\,m^2 = 10\,000\,cm^2$

Fill in the missing numbers.

2.2 kilograms means [ ] grams.

7.6 kilometres means [ ] metres.

£4.58 means [ ] pounds [ ] pence.

0.85 metres means [ ] centimetres.

1.6 tonnes means [ ] kilograms.

0.09 square metres means [ ] square centimetres.

Time spent: _____ min _____ sec.   Total: _____ out of 18

Understanding decimals (1)

Date: _____
Day of Week: _____

STEP 1 — Warm-up (1 min)

Represent the shaded parts as a fraction and a decimal.

1.   ☐ ☐

2.  ☐ ☐

3.   ☐ ☐

4. 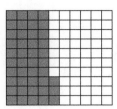 ☐ ☐

STEP 2 — Rapid calculation (2.5 min)

Fill in the missing fractions and decimals.

1.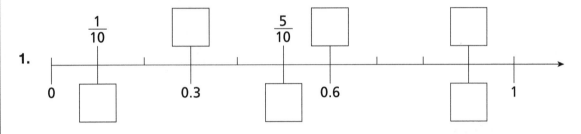

2.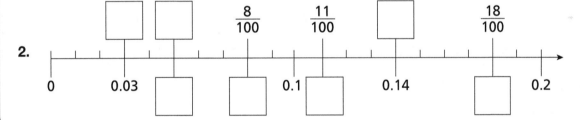

STEP 3 — Challenge (1.5 min)

Answer these. Give your answers as decimals.

$\dfrac{64}{100} - \dfrac{35}{100} =$ ☐

$\dfrac{220}{1000} + \dfrac{498}{1000} =$ ☐

$\dfrac{85}{100} - \dfrac{75}{100} =$ ☐

$\dfrac{745}{1000} - \dfrac{602}{1000} =$ ☐

$\dfrac{9}{10} - \dfrac{3}{10} =$ ☐

$\dfrac{49}{200} + \dfrac{112}{200} =$ ☐

$\dfrac{4}{10} + \dfrac{5}{10} =$ ☐

$\dfrac{7}{100} + \dfrac{5}{100} =$ ☐

Time spent: _____ min _____ sec. Total: _____ out of 29

60 Understanding decimals (2)

STEP 1 Warm-up

Fill in the missing fractions and decimals.

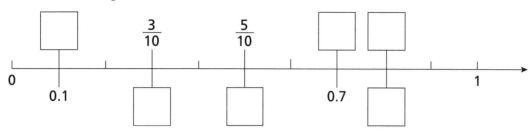

STEP 2 Rapid calculation

TIP Look for the pattern.
$\frac{1}{10} = 0.1$, $\frac{1}{100} = 0.01$ and $\frac{1}{1000} = 0.001$

1. Convert the fractions to decimals.

   $\frac{2}{10} = \square$ $\frac{15}{100} = \square$ $\frac{24}{100} = \square$

   $\frac{136}{100} = \square$ $\frac{97}{100} = \square$ $\frac{8}{100} = \square$

2. Convert the decimals to fractions.

   $0.6 = \frac{\square}{\square}$ $0.28 = \frac{\square}{\square}$ $0.04 = \frac{\square}{\square}$ $1.28 = \frac{\square}{\square}$ $0.07 = \frac{\square}{\square}$

STEP 3 Challenge

Answer these. Give your answers as decimals.

Three $\frac{1}{10}$s are ☐.   Eighteen 0.01s are ☐.

Five 0.01s are ☐.   Nine $\frac{1}{100}$s are ☐.

Twelve $\frac{1}{100}$s are ☐.   Four $\frac{1}{10}$s are ☐.

Ninety-one 0.01s are ☐.   Two hundred and seventy 0.01s are ☐.

Understanding decimals (3)

STEP 1 — Warm-up

Convert the fractions to decimals and convert the decimals to fractions.

$\frac{8}{10}$ = ☐ $\frac{72}{100}$ = ☐ $\frac{36}{100}$ = ☐ $\frac{81}{100}$ = ☐ $\frac{193}{100}$ = ☐

0.7 = ☐   0.57 = ☐   0.02 = ☐   3.7 = ☐   0.35 = ☐

STEP 2 — Rapid calculation

Complete these sentences.

1. One 0.1 is ☐ ; two 0.1s are ☐ ; five 0.1s are ☐ ; ten 0.1s are ☐ .

2. One 0.01 is ☐ ; five 0.01s are ☐ ; ten 0.01s are ☐ .

3. Twenty-three 0.01s are ☐ ; two hundred and thirty 0.01s are ☐ .

4. One hundred 0.01s are ☐ ; one thousand 0.01s are ☐ .

5. Ninety-nine 0.01s are ☐ ; ninety-nine 0.1s are ☐ .

6. The number which consists of three hundred and sixty 0.01s is ☐ .

7. .... 0.01s are 0.92 and nineteen ☐ are 1.9.

STEP 3 — Challenge

Fill in the missing numbers.

1. 0.5 consists of ☐ 0.1s; 0.3 consists of ☐ 0.1s.

2. 0.53 consists of ☐ 0.01s; 0.09 consists of ☐ 0.01s; 0.82 consists of ☐ 0.01s.

3. 0.07 consists of ☐ 0.01s; 0.39 consists of ☐ 0.01s; 0.77 consists of ☐ 0.01s.

4. 0.9 consists of ☐ 0.1s or ☐ 0.01s.

5. Two hundred and ten 0.01s are ☐ .

62 Understanding decimals (4)

Date: _____
Day of Week: _____

STEP 1 Warm-up (1 min)

Write these decimals in digits.

Zero point two five	
One hundred point seven	
Seven point four two	
Zero point zero five	

Thirty-one point three one	
Six point five four	
Sixty-three point one eight	
Forty point zero nine	

STEP 2 Rapid calculation (2.5 min)

Fill in the missing numbers.

1. 1.08 consists of ☐ 1s, ☐ 0.1s and ☐ 0.01s. So 1.08 consists of ☐ 0.01s.

2. 61.52 consists of ☐ 10s, ☐ 1s, ☐ 0.1s and ☐ 0.01s. So 61.52 consists of ☐ 0.01s.

3. 5.7 consists of ☐ 1s and ☐ 0.1s. So 5.7 consists of ☐ 0.1s.

4. The decimal that consists of three 1s and one 0.1 is ☐.

5. The decimal that consists of four 100s, eight 10s, five 1s, nine 0.1s and six 0.01s is ☐.

6. 0.8 consists of ☐ 0.1s. So 0.8 consists of ☐ 0.01s.

7. The decimal that consists of twenty-five 0.01s is ☐.

8. The number that consists of thirty 0.1s is ☐.

STEP 3 Challenge (1.5 min)

Complete these sentences.

98.68 consists of ....

89.12 consists of ....

100.78 consists of ....

0.72 consists of ....

7.05 consists of ....

Time spent: _____ min _____ sec. Total: _____ out of 31

Understanding decimals (5)

STEP 1 — Warm-up

Draw lines to match the equivalent decimals.

zero point eight five	34.8
thirty-four point zero eight	0.05
zero point zero five	0.85
thirty-four point eight	34.08

STEP 2 — Rapid calculation

Complete the tables. The first answer has been done for you.

	… is read as
3.72	three point seven two
1.05	
100.01	
3.52	
0.02	

	… is read as
0.01	
51.45	
9.02	
8.07	
0.06	

STEP 3 — Challenge

Write these decimals in digits.

ninety point five ☐   zero point one three ☐

twenty point eight zero ☐   zero point zero two ☐

zero point eight seven ☐   six point zero nine ☐

twelve point zero seven ☐   ten point zero one ☐

Time spent: ____ min ____ sec. Total: ____ out of 21

64 Understanding decimals (6)

Date: _____
Day of Week: _____

STEP 1 Warm-up

Complete the table.

	as a fraction of a metre	expressed in metres as a decimal
5 cm	_____ m	_____ m
70 cm	_____ m	_____ m
29 cm	_____ m	_____ m

STEP 2 Rapid calculation

Write the decimals in digits to complete the tables.

	... is read as
....	one hundred and one point zero one
....	twenty-five point three four
....	three hundred and eighty-two point zero three
....	three point zero four
....	one thousand, two hundred point zero five

	... is read as
....	fifty-one point one five
....	zero point zero five
....	one point zero two
....	ten point zero six
....	three hundred and eighty-three point zero

STEP 3 Challenge

Complete these sentences.

103.15 is read as ....

0.05 is read as ....

8.00 is read as ....

1006.15 is read as ....

70.09 is read as ....

40.38 is read as ....

42.19 is read as ....

7.09 is read as ....

Comparing decimals (1)

STEP 1 — Warm-up (1 min)

Fill in the boxes with >, < or =.

0.87 ☐ 0.78   12.89 ☐ 12.98   5.67 ☐ 5.63   2.08 ☐ 4.23

7.08 ☐ 8.04   11.86 ☐ 10.86   4.2 ☐ 4.13   52.04 ☐ 52.43

STEP 2 — Rapid calculation (2.5 min)

TIP: *To compare two decimals, compare the whole parts first. The number with a greater whole part is greater. If the whole parts are the same, compare the tenths place, then the hundredths … until you find a greater digit with the same place value.*

Fill in the boxes with >, < or =.

9.18 ☐ 8.19   3.19 ☐ 3.12   10.01 ☐ 10.0

5.47 ☐ 5.46   8.34 ☐ 8.34   9.27 ☐ 8.01

11.91 ☐ 12.01   1.14 ☐ 1.11   6.00 ☐ 6

6.0 ☐ 6.01   5.10 ☐ 5.1   1.25 ☐ 1.52

7.86 ☐ 7.68   0.35 ☐ 0.04   58.2 ☐ 59.2

4.1 ☐ 4.01   9.99 ☐ 9.9   7.77 ☐ 7.77

STEP 3 — Challenge (1.5 min)

Fill in the boxes with >, < or =.

1.7 ☐ 1.07   2.4 ☐ 2.41   0.03 ☐ 0.13   4.67 ☐ 4.86

2.03 ☐ 2.04   1.7 ☐ 1.70   6.04 ☐ 6.040   7.87 ☐ 8.78

0.5 ☐ 0.05   1.0 ☐ 1

66 Comparing decimals (2)

Date: _____
Day of Week: _____

STEP 1 — Warm-up (1 min)

Fill in the boxes with >, < or =.

0.25 ☐ 0.52   30.2 ☐ 3.02   7.63 ☐ 6.37   0.85 ☐ 0.58

7.88 ☐ 7.87   10.5 ☐ 10.6   3.9 ☐ 3.82   34.81 ☐ 34.85

STEP 2 — Rapid calculation (2.5 min)

Fill in the boxes with >, < or =.

2.7 ☐ 7.2   5.16 ☐ 51.6   9.01 ☐ 10

6.47 ☐ 7.46   83.47 ☐ 83.43   9.05 ☐ 8.95

10.91 ☐ 10.01 2.14 ☐ 2.41   7.8 ☐ 8

0.01 ☐ 1.01   2.06 ☐ 6.02   1.75 ☐ 1.57

6.58 ☐ 6.85   0.93 ☐ 0.99   86.4 ☐ 84.6

1.1 ☐ 1.10   0.30 ☐ 0.35   10.1 ☐ 10

STEP 3 — Challenge (1.5 min)

Fill in the boxes with >, < or =.

2.6 ☐ 2.06   3.4 ☐ 3.40   0.08 ☐ 0.8   2.67 ☐ 20.67

3.03 ☐ 3.04   5.7 ☐ 5.70   8.5 ☐ 8.52   3.07 ☐ 30.7

0.06 ☐ 0.5   2.03 ☐ 2.30

Time spent: _____ min _____ sec. Total: _____ out of 36

Date: _____
Day of Week: _____

Practice (1) 67

STEP 1 Warm-up (1 min)

Start the timer

Write these fractions as decimals.

$\dfrac{7}{10}$ = ☐ $\dfrac{3}{100}$ = ☐ $\dfrac{38}{100}$ = ☐ $3\dfrac{92}{100}$ = ☐

$\dfrac{9}{100}$ = ☐ $\dfrac{21}{100}$ = ☐ $\dfrac{70}{100}$ = ☐ $\dfrac{5}{100}$ = ☐

$\dfrac{52}{100}$ = ☐ $5\dfrac{42}{100}$ = ☐

STEP 2 Rapid calculation (2.5 min)

Start the timer

Fill in the boxes with >, < or =.

30.8 ☐ 31.1   3.90 ☐ 3.92   0.09 ☐ 0.89   0.80 ☐ 0.84

5.82 ☐ 5.80   90.08 ☐ 90.080   2.12 ☐ 2.21   3.7 ☐ 3.71

0.3 ☐ 0.33   0.34 ☐ 0.340   £0.10 ☐ £0.09   5.05 ☐ 5.49

0.7 ☐ 0.08   2.7 ☐ 2.72   1.2 ☐ 12.01   4 ☐ 4.00

STEP 3 Challenge (1.5 min)

Start the timer

Order these numbers from **least** to **greatest**.

1. 5.05   5.5   5.55   5.0   50

    ....

2. 3.3   3   0.33   33   3.03

    ....

Time spent: _____ min _____ sec. Total: _____ out of 36

68 Practice (2)

STEP 1 Warm-up

Simplify the decimals. The first one has been done for you.

2.10 = **2.1**   70.00 =   0.020 =   90.00 =   7.040 =

54.010 =   17.300 =   0.080 =   0.70 =   9.230 =

STEP 2 Rapid calculation

Complete the sentences.

1. 30 cm equals □/□ m, written in decimal form as □ m.

2. 35 g equals □/□ kg, written in decimal form as □ kg.

3. Two hundred and twenty-five $\frac{1}{100}$s can be written as the mixed number □ and in decimal form as □.

4. There are 9 .... in 0.9; there are 90 .... in 0.90.

5. The two decimals nearest to 0.5 with two decimal places are □ and □.

6. The two decimals nearest to 2 with two decimal places are □ and □.

STEP 3 Challenge

Compare the decimal numbers using the symbols >, < or =.

0.2 □ 0.02   0.28 □ 2.18   0.050 □ 0.05   8.54 □ 8.49

4.47 □ 4.09   85.58 □ 85.85   3.200 □ 3.20   0.78 □ 0.8

Time spent: _____ min _____ sec. Total: _____ out of 29

Date: _____
Day of Week: _____

Acute angles and obtuse angles **69**

STEP 1 — Warm-up

Write the number of each angle in the correct box.

Right angles	Acute angles	Obtuse angles

STEP 2 — Rapid calculation

Which type of angle do the clock hands form at these times?

9:30   10:00   7:45   5:30   3:00   1:30   11:30   2:00   9:00

Write each time in the correct box.

Right angles	Acute angles	Obtuse angles

STEP 3 — Challenge

Which type of angles are shown in the diagram? The first one has been done for you.

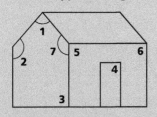

∠1: **acute** angle   ∠2: _____ angle   ∠3: _____ angle

∠4: _____ angle   ∠5: _____ angle   ∠6: _____ angle

∠7: _____ angle

©HarperCollinsPublishers 2019   Time spent: ____ min ____ sec. Total: ____ out of 25

70 Triangles, quadrilaterals and pentagons

Date: _____
Day of Week: _____

STEP 1 Warm-up

Complete these descriptions.

1. A shape enclosed by three .... is called a ....

2. A shape enclosed by .... line segments is called a quadrilateral.

3. A shape enclosed by five .... is called a ....

STEP 2 Rapid calculation

Sort these shapes by writing their numbers in the correct boxes.

 ①   ②   ③   ④   ⑤

 ⑥   ⑦   ⑧    ⑨

Triangles	Quadrilaterals	Pentagons	Other

STEP 3 Challenge

Complete the table.

Shape	Number of sides	Number of angles	Name
△			
▱			
⏢			
⬠			

Time spent: _____ min _____ sec. Total: _____ out of 26

Classification of triangles 71

STEP 1 Warm-up

Complete these descriptions.

1. A triangle can be classified using its angles as a(n) .... triangle, a(n) .... triangle or a(n) .... triangle.
2. A triangle with an obtuse angle is called a(n) .... triangle.
3. A triangle with three acute angles is called a(n) .... triangle.
4. A triangle with a right angle is called a(n) .... triangle.
5. A triangle can be classified using its sides as a(n) .... triangle, a(n) .... triangle or a(n) .... triangle.

STEP 2 Rapid calculation

Sort these triangles by writing their numbers in the correct box(es).

Acute	Obtuse	Right-angled	Isosceles	Equilateral

STEP 3 Challenge

Complete these sentences.

1. A triangle has at least ☐ acute angle(s).
2. The side lengths of a triangle are 3 cm, 3 cm and 4 cm. According to its sides it is a(n) .... triangle.
3. A triangle with one axis of symmetry is called a(n) .... triangle.

    A triangle with three axes of symmetry is called a(n) .... triangle.
4. A triangle with three equal .... is called an equilateral triangle.

    It also has three equal ....

72 Line symmetry

Date: _____
Day of Week: _____

STEP 1 Warm-up

Tick the boxes next to the symmetrical figures, then draw the axes of symmetry.

1. ☐
2. ☐
3. ☐
4. ☐
5. ☐
6. ☐

STEP 2 Rapid calculation

Tick the boxes next to the figures with line symmetry, then draw the axes of symmetry.

1. ☐
2. ☐
3. ☐
4. ☐

5. 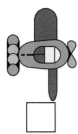 ☐
6. ☐
7. ☐
8. ☐

STEP 3 Challenge

Complete these sentences.

1. A rectangle has ☐ axes of symmetry and a square has ☐ axes of symmetry.

2. An equilateral triangle has ☐ axes of symmetry.

3. An isosceles (but not equilateral) triangle has ☐ axes of symmetry.

4. A circle has .... axes of symmetry.

Time spent: _____ min _____ sec. Total: _____ out of 19

Areas of rectangles and squares (1)

STEP 1 — Warm-up

Complete these.

1. Area of a rectangle = ... × ....

    The area of the rectangle shown is [ ] cm².

2. Area of a square = ... × ....

    The area of the square shown is [ ] cm².

STEP 2 — Rapid calculation

1. Work out the areas of these rectangles.

    length 7 cm, width 5 cm, area = [ ] cm²   length 20 cm, width 15 cm, area = [ ] cm²

    length 50 cm, width 35 cm, area = [ ] cm²   length 70 m, width 42 m, area = [ ] m²

    length 60 m, width 45 m, area = [ ] m²   length 40 cm, width 20 cm, area = [ ] cm²

2. Work out the areas of these squares.

    side length 15 cm, area = [ ] cm²   side length 80 cm, area = [ ] cm²

    side length 11 m, area = [ ] m²   side length 20 cm, area = [ ] cm²

STEP 3 — Challenge

1. Work out the areas of these rectangles.

    length 7 m, width 50 cm, area = [ ] cm²   length 1 m, width 32 cm, area = [ ] cm²

    length 2 m, width 40 cm, area = [ ] cm²

2. Work out the areas of these squares.

    side length 5 m, area = [ ] cm²   side length 7 m, area = [ ] cm²

    side length 4 m, area = [ ] cm²

74 Areas of rectangles and squares (2)

Date: _____
Day of Week: _____

STEP 1 Warm-up

Complete these sentences. Use abbreviated metric units.

1. A carpet is about 8 .... in length and 5 .... in width. Its area is about [ ].

2. A door is about 200 .... in length and 100 .... in width. Its area is about [ ].

3. A photo is about 12 .... in length and 5 .... in width. Its area is about [ ].

4. A sports pitch is about 100 .... in length and 50 .... in width. Its area is about [ ].

STEP 2 Rapid calculation

1. Work out the areas of these rectangles.

   length 52 m, width 20 m, area = [ ] m²   length 5 cm, width 4 cm, area = [ ] cm²

   length 16 cm, width 10 cm, area = [ ] cm²   length 24 m, width 20 m, area = [ ] m²

   length 45 m, width 30 m, area = [ ] m²   length 30 cm, width 14 cm, area = [ ] cm²

2. Work out the areas of these squares.

   side length 12 cm, area = [ ] cm²   side length 90 cm, area = [ ] cm²

   side length 16 m, area = [ ] m²   side length 40 cm, area = [ ] cm²

STEP 3 Challenge

1. Find the missing side in each rectangle.

   length 20 cm, area 240 cm², width [ ] cm

   width 14 cm, area 560 cm², length [ ] cm

   length 30 cm, area 480 cm², width [ ] cm

2. Find the missing side in each square.

   area 4900 cm², side length [ ] cm

   area 8100 cm², side length [ ] cm

   area 144 cm², side length [ ] cm

Time spent: _____ min _____ sec. Total: _____ out of 24

Areas of rectangles and squares (3)

STEP 1 Warm-up

Work out the area of each shape.

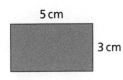 5 cm, 3 cm → ☐ cm²

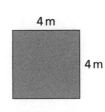 4 m, 4 m → ☐ m²

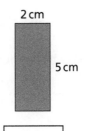 2 cm, 5 cm → ☐ cm²

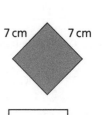 7 cm, 7 cm → ☐ cm²

STEP 2 Rapid calculation

Complete the tables.

Rectangle	Length	8 cm	30 m	25 cm	40 cm	
	Width	4 cm	12 m			20 m
	Area			500 cm²	960 cm²	360 m²

Square	Side length	9 cm	14 m			
	Area			36 cm²	100 cm²	196 cm²

STEP 3 Challenge

1. Work out the missing side in each rectangle.

   length 35 m, area 350 m², width ☐ m

   width 40 m, area 2400 m², length ☐ m

   length 18 m, area 216 m², width ☐ m

   width 20 m, area 640 m², length ☐ m

2. Work out the missing side in each square.

   area 225 m², side length ☐ m

   area 400 m², side length ☐ m

   area 144 m², side length ☐ m

   area 324 m², side length ☐ m

76 Converting between kilometres and metres

STEP 1 Warm-up

Which is the most suitable metric unit? Fill in the abbreviations.

1. A baby giraffe is about 2 .... in height.
2. The distance from Mike's home to the library is 5 ....
3. For a running test in the PE class, every pupil has to run 100 ....
4. The full marathon distance is about 42 ....
5. A bicycle travels at around 300 .... a minute.
6. One lap of a running track is 400 m; an athlete runs five laps, which is 2 ....
7. A plane flies 900 .... an hour.
8. Mount Everest is about 8848 .... in height.

STEP 2 Rapid calculation

Complete these conversions.

1 km = ☐ m   4 km = ☐ m   30 000 m = ☐ km
9 km = ☐ m   5000 m = ☐ km   18 000 m = ☐ km
7000 m = ☐ km   80 km = ☐ m   38 000 m = ☐ km
19 km = ☐ m   8 km 750 m = ☐ m   7 km 7 m = ☐ m
99 000 m = ☐ km   16 km 40 m = ☐ m   3 km 450 m = ☐ m

STEP 3 Challenge

Fill in the missing numbers.

5 km 20 m = ☐ m   9 km 67 m = ☐ m   300 m + 4 km = ☐ m
2 km + 5800 m = ☐ m   8 km − 430 m = ☐ m   8 km − 80 m = ☐ m
30 km − ☐ m = 600 m   ☐ km − 1500 m = 500 m

Perimeters of rectangles and squares (1)

STEP 1 — Warm-up

1. The length of the outline of a plane figure is called the ....

2. Perimeter of a rectangle = ☐ × .... + ☐ × ....

3. Perimeter of a square = ☐ × ....

4. Work out the perimeter of each shape.

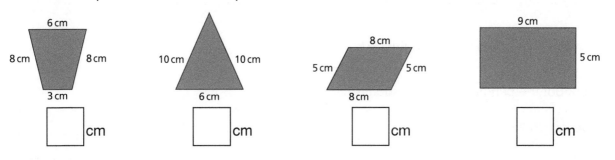

STEP 2 — Rapid calculation

Complete the tables.

Rectangles	Length	18 cm	25 cm	40 cm	50 cm	100 cm
	Width	10 cm	20 cm	35 cm	42 cm	80 cm
	Perimeter					

Squares	Side length	34 cm	50 cm	72 cm	88 cm	43 cm
	Perimeter					

STEP 3 — Challenge

1. Find the missing side in each rectangle.

   perimeter 48 cm, length 8 cm, width ☐ cm

   perimeter 60 cm, width 4 cm, length ☐ cm

   perimeter 96 cm, length 36 cm, width ☐ cm

   perimeter 100 cm, width 18 cm, length ☐ cm

2. Find the missing side in each square.

   perimeter 76 cm, side length ☐ cm

   perimeter 92 cm, side length ☐ cm

   perimeter 64 cm, side length ☐ cm

   perimeter 48 cm, side length ☐ cm

78 Perimeters of rectangles and squares (2)

Date: _____
Day of Week: _____

STEP 1 — Warm-up

1. Work out the perimeter of each rectangle.

   length 7 cm, width 4 cm, perimeter ☐ cm

   length 12 cm, width 8 cm, perimeter ☐ cm

   length 24 cm, width 17 cm, perimeter ☐ cm

   length 43 cm, width 35 cm, perimeter ☐ cm

2. Work out the perimeter of each square.

   side length 24 cm, perimeter ☐ cm   side length 18 cm, perimeter ☐ cm

   side length 60 cm, perimeter ☐ cm   side length 94 cm, perimeter ☐ cm

STEP 2 — Rapid calculation

Complete the tables.

Rectangles	Length	24 cm		18 m	7 m	
	Width		20 cm	5 m		12 m
	Perimeter	78 cm	120 cm		24 m	46 m

Squares	Side length	23 cm			15 cm	
	Perimeter		72 m	32 m		96 m

STEP 3 — Challenge

Fill in the missing numbers.

1. The area of a rectangle is 1200 cm² and its length is 60 cm. Its perimeter is ☐ cm.

2. The perimeter of a rectangle is 88 m and its width is 10 m. Its area is ☐ m².

3. The area of a square is 49 cm². Its perimeter is ☐ cm.

4. The perimeter of a square is 48 cm. Its area is ☐ cm².

Time spent: _____ min _____ sec. Total: _____ out of 22

Solving problems involving time and money

STEP 1 Warm-up

Fill in the missing numbers.

1 h = ☐ min   1 min = ☐ sec

1 day = ☐ h   1 week = ☐ days

1 year = ☐ months   1 h = ☐ sec

£1 = ☐ pence   2 min = ☐ sec

STEP 2 Rapid calculation

Fill in the missing numbers.

4 h = ☐ min   3.5 min = ☐ sec   75 min = ☐ h

1.5 days = ☐ h $\frac{2}{5}$ h = ☐ min $\frac{2}{3}$ days = ☐ h

$\frac{1}{2}$ year = ☐ months $\frac{4}{5}$ of £1 = ☐ pence   £9.50 = ☐ pence

70 pence = £ ☐   2 weeks = ☐ days   608 pence = £ ☐

STEP 3 Challenge

Fill in the missing numbers.

280 h = ☐ days ☐ h   590 pence = £ ☐ and ☐ pence

$\frac{3}{5}$ h = ☐ min = ☐ sec   ☐ h = 45 min = ☐ sec

4 weeks = ☐ days = ☐ h   2.5 years = ☐ months

80 Solving calculations in steps

Date: _____
Day of Week: _____

STEP 1 — Warm-up (1 min)

Answer these.

600 ÷ 40 = ☐   180 + 52 = ☐   620 − 380 = ☐   800 ÷ 25 = ☐

99 × 7 = ☐   424 ÷ 4 = ☐   483 + 278 = ☐   37 × 40 = ☐

STEP 2 — Rapid calculation (2.5 min)

Write a number sentence with brackets for each tree diagram. The first one has been done for you.

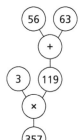
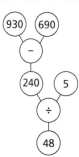
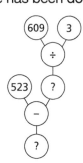

3 × (56 + 63) = 357

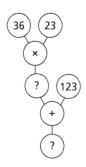
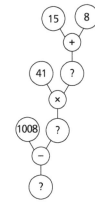
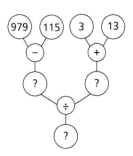

STEP 3 — Challenge (1.5 min)

Answer these. Remember to complete the calculations in brackets first.

112 − (125 ÷ 5) = ☐   (228 + 129) ÷ 7 = ☐   70 × (53 − 38) = ☐

(42 × 8) − 138 = ☐   (675 − 451) ÷ 4 = ☐   160 − (160 ÷ 4) = ☐

(1250 − 250) ÷ 25 = ☐   105 × (43 − 37) = ☐   (52 × 40) − 96 = ☐

Time spent: ____ min ____ sec. Total: ____ out of 22

ANSWERS

Test 1
Step 1:
From top left, left to right: 33; 418; 514; 657; 446; 373; 188; 199; 1070; 801

Step 2:
From top left, left to right: 1030; 100; 1090; 1300; 2400; 1000; 5650; 7460; 1300; 8309; 4781; 6077; 5356; 2400; 10 720

Step 3:
From top left, left to right: 100; 1000; 100; 10; 1000; 100; 1000; 10

Test 2
Step 1:
First row: seven thousand three hundred and twenty-six;
three thousand eight hundred and twenty-one;
eight thousand two hundred and thirty;
nine thousand five hundred and twenty-six
Second row: nine thousand six hundred and sixty-two;
five thousand six hundred;
three thousand seven hundred and two;
eight thousand seven hundred and fourteen

Step 2:
From top left, left to right:
7000 + 700 + 80 + 8;
1000 + 500 + 60 + 0;
2000 + 800 + 0 + 0;
3000 + 300 + 90 + 5;
9000 + 0 + 60 + 5;
4000 + 500 + 80 + 6;
5000 + 0 + 80 + 0;
3000 + 400 + 90 + 9;
7000 + 0 + 80 + 6;
8000 + 500 + 50 + 0;
6000 + 0 + 30 + 4

Step 3:
From top: 2100; 9850; 4626; 7003; 5030; 4106; 6820; 9400

Test 3
Step 1:
From top left, left to right: 4385; 6054; 2497; 8014; 9300; 7690; 3900; 5005

Step 2:
From top left, left to right: 4980; 5328; 6320; 7085; 3030; 7891; 2095; 3957; 4806; 6853; 5459; 2005

Step 3:
1. From left to right: 8990, 9000; 5770, 5780
2. From left to right: 6300, 6400; 2700, 2800; 9700, 9800
3. From left to right: 4000, 5000; 6000, 7000; 8000, 9000

Test 4
Step 1:
1. 3900, 3910; 5830, 5840; 4400, 4410
2. 3200, 3300; 4500, 4600; 5300, 5400
3. 9000, 10000; 8000, 9000; 2000, 3000

Step 2:
First row: 4970; 3020; 2390; 8950; 9310
Second row: 3400; 4700; 7400; 6500; 5600
Third row: 5000; 3000; 3000; 4000; 6000

Step 3:
From top left, left to right:
2300, 2350; 7485, 7385; 5753, 3753; 3256, 3356; 515, 715; 7100, 8100; 3812, 3813; 4350, 4460

Test 5
Step 1:
From top left, left to right: 1; 5; 10; 50; 100; 500; 1000

Step 2:
From top left, left to right: 8; 4; 12; 15; 43; 66; 9; 82; 37

Step 3:
From top left, left to right: >; =; <; >; <; =

Test 6
Step 1:
From top left, left to right: 700; 750; 990; 500; 920; 710; 634; 679

Step 2:
From top left, left to right: 715; 730; 936; 925; 688; 689; 878; 599; 998; 597; 786; 499; 682; 928; 909; 970; 802; 786; 964; 1053

Step 3:
From top left, left to right: 1251; 1254; 1353; 996; 1225; 1292; 1715; 1934

Test 7
Step 1:
From top left, left to right: 950; 920; 770; 900; 900; 920; 774; 897

Step 2:
From top left, left to right: 797; 879; 900; 889; 592; 801; 891; 582; 947; 1001; 692; 888; 790; 872; 841; 981; 931; 910; 964; 1041

Step 3:
From top left, left to right: 1114; 1243; 1313; 1291; 1483; 1554; 1625; 1952

©HarperCollins*Publishers* 2019

Answers

Test 8
Step 1:
From top left, left to right: 35; 26; 45; 36; 19; 39; 29; 39; 45; 19

Step 2:
From top left, left to right: 222; 122; 313; 243; 431; 503; 422; 222; 338; 319; 227; 415; 458; 128; 279; 579; 141; 185

Step 3:
From top left, left to right: <; >; =; <; >; =; >; <; >

Test 9
Step 1:
From top left, left to right: 28; 8; 37; 28; 18; 48; 19; 39; 18; 39

Step 2:
From top left, left to right: 109; 218; 219; 217; 489; 515; 216; 327; 392; 277; 358; 89; 289; 268; 377; 378; 134; 212

Step 3:
From top left, left to right: 184; 378; 279; 494; 368; 268; 287; 189

Test 10
Step 1:
From top left, left to right: 210; 320; 950; 920; 790; 980; 290; 120; 240; 480

Step 2:
From top left, left to right: 117; 859; 829; 210; 605; 505; 295; 610; 395; 140; 402; 896; 892; 672; 500; 176

Step 3:
From top left, left to right: 210; 575; 220; 166; 670; 340; 545; 870; 575

Test 11
Step 1:
From top left, left to right: 1080; 460; 700; 540; 270; 930; 320; 810; 790; 280

Step 2:
From top left, left to right: 895; 460; 594; 330; 500; 510; 796; 640; 990; 120; 205; 413; 330; 782; 589

Step 3:
From top left, left to right: 336; 568; 723; 536; 588; 673; 856; 360; 123

Test 12
Step 1:
From top left, left to right: 18; 19; 74; 16; 25; 91; 72; 9; 31; 92

Step 2:
From top left, left to right: 666; 413; 211; 888; 688; 113; 427; 227; 337; 890; 219; 771; 664; 426; 677; 865; 107; 693; 983; 249

Step 3:
From top left, left to right: 466; 682; 269; 248; 690; 806; 317; 249; 611

Test 13
Step 1:
From top left, left to right: 80; 28; 52; 54; 91; 37; 790; 530; 260; 800

Step 2:
From top left, left to right: 48; 28; 110; 120; 770; 75; 85; 162; 60; 680; 520; 48; 250; 1000; 309

Step 3:
From top left, left to right: 100; 250; 160; 170; 22; 165; 1000; 21

Test 14
Step 1:
From top left, left to right: 970; 140; 800; 3680; 130; 94; 350; 3200; 840; 46

Step 2:
From top left, left to right: 53; 70; 940; 730; 710; 40; 390; 320; 1000; 59; 231; 400; 130; 176; 590

Step 3:
From top left, left to right: 90; 420; 610; 156; 402; 400; 260; 40

Test 15
Step 1:
From top left, left to right: 4600; 3400; 5900; 9100; 6900; 8400; 5600; 1700; 4400; 6500

Step 2:
First table, from top: 6700, 6692; 8600, 8594; 7800, 7793; 8000, 8032
Second table, from top: 2900, 2882; 4000, 4069; 6200, 6278; 3200, 3177

Step 3:
First table, from top: 4800, 4775; 9300, 9262; 2800, 2779
Second table, from top: 8300, 8271; 4800, 4877; 9300, 9262

Test 16
Step 1:
From top left, left to right: 500; 4500; 6000; 1300; 3900; 9000; 9000; 8000; 9000

Answers

Step 2:
From top left, left to right: 4900; 3900; 8400; 8900; 9700; 6800; 7700; 9900; 5900; 5800; 3600; 8800; 6800; 9400; 8700

Step 3:
From top left, left to right: 3406; 5744; 6577; 5559; 5861; 5899; 8927; 4028

Test 17
Step 1:
From top left, left to right: 570; 594; 860; 919; 837; 660; 580; 575; 558; 575

Step 2:
From top left, left to right: 3600; 2801; 6900; 7138; 9900; 6800; 5906; 4794; 8785; 3965; 8207; 8555

Step 3:
From top left, left to right: 9901; 7783; 7280; 7693; 8602; 6618; 8904; 7321

Test 18
Step 1:
From top left, left to right: 200; 850; 200; 300; 500; 600; 100; 300; 300; 600

Step 2:
From top left, left to right: 2100; 3500; 4100; 3300; 1500; 3000; 5400; 3200; 6000; 2400; 2600; 3300; 2200; 4400; 1800

Step 3:
From top left, left to right: 3101; 5130; 3118; 2060; 4224; 2030; 2402; 3223

Test 19
Step 1:
From top left, left to right: 2000; 5000; 5000; 3900; 3500; 4000; 3000; 7000; 1500; 3600

Step 2:
From top left, left to right: 1522; 3444; 3588; 3318; 894; 1039; 2063; 1896; 1248; 2337; 2452; 2239

Step 3:
From top left, left to right: 2000; 2105; 5132; 2319; 2200; 5502; 2438; 1340

Test 20
Step 1:
From top left, left to right: 0; 0; 8; 79; 11; 10; 2; 6; 12; 60; 63; 48

Step 2:
From top left, left to right: 12; 48; 18; 66; 30; 72; 42; 54; 4; 5; 3; 9; 8; 60; 1; 54; 11; 6; 36; 10

Step 3:
From top left, left to right: 45; 60; 74; 59; 90; 11; 19; 75; 93

Test 21
Step 1:
From top left, left to right: 4; 78; 5; 7; 88; 18; 3; 72; 9; 72; 83; 12

Step 2:
From top left, left to right: 60; 7; 11; 54; 18; 6; 42; 1; 6; 6; 6; 9; 12; 10; 0; 0; 6; 6; 0; 4

Step 3:
From top left, left to right: 18; 6; 6; 24; 6; 54; 9; 6; 66

Test 22
Step 1:
From top left, left to right: 32; 44; 8; 28; 9; 44; 30; 63; 54; 48; 71; 24

Step 2:
From top left, left to right: 14; 28; 56; 35; 42; 14; 21; 28; 35; 84; 56; 63; 7; 70; 21; 0; 77; 42; 84; 49

Step 3:
From top left, left to right: 3; 5; 7; 2; 8; 11; 12; 6; 9

Test 23
Step 1:
From top left, left to right: 30; 19; 8; 33; 8; 39; 45; 36; 60; 28; 45; 48

Step 2:
From top left, left to right: 21; 35; 49; 42; 77; 0; 63; 84; 7; 7; 8; 7; 11; 10; 9; 0; 7; 7; 7; 12

Step 3:
From top left, left to right: 10; 45; 27; 47; 23; 48; 69; 41; 70

Mind Gym:
1. 2, 6
2. 6, 3

Test 24
Step 1:
From top left, left to right: 48; 65; 5; 9; 9; 20; 18; 72; 12; 35; 72; 44

Step 2:
From top left, left to right: 49; 42; 63; 35; 0; 77; 2; 84; 6; 7; 56; 3; 0; 7; 8; 11; 7; 7; 12; 7

Step 3:
From top left, left to right: 28; 7; 96; 11; 3; 49; 7; 7; 0

Answers

Test 25
Step 1:
From top left, left to right: 33; 18; 0; 81; 84; 72; 11; 22; 21; 0; 28; 12

Step 2:
From top left, left to right: 18; 45; 81; 90; 54; 99; 63; 72; 36; 108; 0; 6; 2; 12; 9; 11; 7; 3; 5; 10

Step 3:
From top left, left to right: 27; 9; 5; 63; 9; 90; 4; 9; 6

Test 26
Step 1:
From top left, left to right: 36; 11; 44; 80; 81; 99; 5; 84; 39; 9; 44; 9

Step 2:
From top left, left to right: 0; 45; 90; 4; 54; 3; 81; 1; 7; 36; 5; 10; 9; 11; 9; 0; 9; 63; 9; 108

Step 3:
From top left, left to right: =; <; =; <; <; <; >; =; <

Test 27
Step 1:
From top left, left to right: 158; 80; 960; 30; 14; 600; 14; 136; 70

Step 2:
From top left, left to right: 288; 190; 264; 52; 105; 16; 174; 110; 130; 325; 28; 32; 70; 434; 23

Step 3:
From top left, left to right: 91; 600; 616; 112; 81; 42; 9; 7; 8

Test 28
Step 1:
From top left, left to right: 68; 26; 106; 372; 36; 0; 207; 125; 150

Step 2:
From top left, left to right: 63; 293; 71; 36; 104; 216; 198; 630; 2084; 741; 1509; 1680; 960; 45; 23

Step 3:
From top left, left to right: 48; 50; 848; 153; 825; 120; 73; 392

Test 29
Step 1:
From top left, left to right: 55; 14; 69; 20; 50; 88; 17; 37; 100; 29; 27; 84

Step 2:
From top left, left to right:
1. 8, 16; 9, 27; 10, 40; 6, 30; 9, 54; 6, 42
2. 56; 72; 49; 22; 60; 48

Step 3:
From top:
$10 \times 4 + 9 \times 4 = 40 + 36 = 76$
$10 \times 8 + 2 \times 8 = 80 + 16 = 96$
$10 \times 4 + 3 \times 4 = 40 + 12 = 52$
$10 \times 6 + 5 \times 6 = 60 + 30 = 90$
Other suitable answers are possible.

Test 30
Step 1:
From top left, left to right: 77; 10; 61; 17; 62; 68; 23; 18; 65; 35; 45; 83

Step 2:
From top left, left to right: 2, 4; 4, 12; 3, 12; 2, 10; 6, 6, 36; 4, 7, 28; 5, 5, 25; 5, 9, 45; 4, 7, 28; 12, 8, 96; 11, 3, 33; 12, 9, 108

Step 3:
From top: 6, 24; 12, 5, 60; 11, 7, 77; 10, 9, 90

Mind Gym:
※ = 15   ◎ = 9

Test 31
Step 1:
From top left, left to right:
50; 28; 45; 32; 8; 6; 91; 28; 49; 7; 9; 0

Step 2:
From top left, left to right:
$4 \times 9 = 36$; $2 \times 3 = 6$; $3 \times 8 = 24$; $3 \times 7 = 21$; $3 \times 4 = 12$; $3 \times 3 = 9$; $7 \times 5 = 35$; $5 \times 9 = 45$; $1 \times 4 = 4$; $2 \times 10 = 20$; $12 \times 9 = 108$; $12 \times 7 = 84$

Step 3:
From top left, left to right:
$7 \times 9 = 63$; $11 \times 8 = 88$; $6 \times 6 = 36$; $12 \times 5 = 60$ or $6 \times 10 = 60$

Test 32
Step 1:
From top left, left to right: 100; 170; 325; 129; 1375; 170; 116; 369; 211; 520

Step 2:
From top left, left to right: 225; 669; 780; 328; 296; 222; 1360; 156; 648; 322; 558; 234; 140; 496; 664

Step 3:
From top left, left to right: 2100; 235; 91; 810; 221; 140; 47; 336

Answers

Test 33
Step 1:
From top left, left to right: 14; 125; 299; 75; 365; 648; 11; 250; 41; 368

Step 2:
From top left, left to right: 288; 372; 639; 288; 480; 595; 264; 336; 140; 460; 216; 189; 160; 473; 600

Step 3:
From top left, left to right: 67; 203; 729; 256; 400; 594; 71; 56

Test 34
Step 1:
From top left, left to right: 900; 4200; 2000; 600; 4800; 5400; 2700; 1400; 3200; 2500

Step 2:
From top left, left to right: 480; 940; 480; 570; 1000; 260; 960; 1100; 960; 910; 750; 990; 720; 510; 960

Step 3:
From top left, left to right: 800; 500; 450; 520; 1440; 960; 680; 1280; 1050

Test 35
Step 1:
From top left, left to right: 33; 48; 70; 108; 84; 162; 78; 95; 56; 117

Step 2:
From top left, left to right: 330; 480; 700; 1080; 840; 1620; 780; 320; 940; 360; 760; 820; 570; 700; 640

Step 3:
From top left, left to right: 4050; 1880; 380; 860; 1720; 1100; 1400; 2250

Mind Gym:
1. 642 × 8 = 5136
2. 468 × 2 = 936

Test 36
Step 1:
From top left, left to right: 1800; 5600; 2500; 1800; 1600; 3000; 2100; 1600; 1500; 4800

Step 2:
From top left, left to right: 880; 450; 960; 1180; 1140; 1020; 960; 1190; 480; 1120; 1050; 520; 750; 1300; 1330; 390; 280; 300; 720; 840

Step 3:
From top left, left to right: 920; 1300; 900; 990; 720; 690; 2500; 750; 900; 760; 990; 1640

Test 37
Step 1:
1. 24, 240
2. 72, 720
3. 26, 260
4. 160, 1600
5. 203, 2030

Step 2:
Alternative combinations are possible:
2. 20, 3, 980, 147, 1127
3. 35, 40, 35, 6, 1400, 210, 1610
4. 21, 10, 21, 8, 210, 168, 378
5. 29, 50, 29, 2, 1450, 58, 1508
6. 71, 30, 71, 4, 2130, 284, 2414

Step 3:
From top left, left to right: 165; 275; 418; 891; 374; 154; 693; 1012

Test 38
Step 1:
From top left, left to right: 76; 304; 175; 77; 156; 468; 135; 111

Step 2:
From top, left to right: 10, 3, 370, 111, 481; 20, 5, 320, 80, 400; 10, 8, 410, 328, 738; 38, 10, 38, 2, 380, 76, 456; 84, 30, 84, 1, 2520, 84, 2604; 74, 20, 74, 3, 1480, 222, 1702; 52, 10, 52, 4, 520, 208, 728; 46, 20, 46, 7, 920, 322, 1242; 68, 20, 68, 6, 1360, 408, 1768

Step 3:
From top left, left to right: 814; 858; 506; 693; 594; 539; 759; 946

Mind Gym:
4 − 2 = 2

Test 39
Step 1:
From top left, left to right: 1520; 4320; 720; 980; 1710; 1520; 2190; 1300

Step 2:
From top, left to right: 31, 20, 31, 2, 620, 62, 682;
36, 10, 36, 4, 360, 144, 504;
18, 20, 18, 5, 360, 90, 450;
43, 10, 43, 7, 430, 301, 731;
38, 10, 38, 6, 380, 228, 608;
77, 40, 77, 1, 3080, 77, 3157;
87, 30, 87, 1, 2610, 87, 2697;
64, 20, 64, 5, 1280, 320, 1600;
53, 10, 53, 8, 530, 424, 954;
94, 30, 94, 1, 2820, 94, 2914

Answers

Step 3:
From top left, left to right: 1050; 630; 570; 690; 360; 1080; 990; 1410

Test 40
Step 1:
From top left, left to right: 4; 6; 32; 1700; 3300; 1250; 6720; 910; 3

Step 2:
From top left, left to right: 35; 19; 4140; 1080; 13; 2610; 2250; 12; 19; 18; 22; 1320; 7; 18; 3920

Step 3:
From top left, left to right: 10; 19; 130; 60; 3780; 1500; 30; 24

Test 41
Step 1:
1. 45, 450
2. 140, 1400
3. 128, 1280
4. 216, 2160
5. 108, 1080

Step 2:
From top left, left to right: 960; 1080; 1140; 1290; 2160; 3150; 1280; 2200; 1110; 3200; 1890; 1560; 1120; 1890; 4080

Step 3:
From top left, left to right: 2310; 2200; 870; 2240; 3350; 2070; 6880; 3840

Test 42
Step 1:
1. 96, 960
2. 112, 1120
3. 144, 1440
4. 144, 1440
5. 190, 1900

Step 2:
From top left, left to right: 810; 2680; 2800; 3150; 2880; 3360; 1770; 2850; 1850; 3120; 4480; 2160; 2080; 2970; 4270

Step 3:
From top left, left to right: 2160; 1680; 3250; 2760; 3160; 4640; 2660; 2430

Test 43
Step 1:
From top left, left to right: 200; 40; 120; 50; 90; 60; 80; 80; 60; 30

Step 2:
From top left, left to right: 20; 5; 10; 5; 60; 6; 8; 8; 7; 5; 9; 6; 9; 13; 4

Step 3:
From top left, left to right: 9; 9; 7; 14; 8; 7; 6; 42

Test 44
Step 1:
From top left, left to right: 4; 6; 6; 9; 8; 6; 5; 5; 4

Step 2:
From top left, left to right: 7; 6; 9; 4; 13; 8; 10; 9; 8; 18; 9; 8; 12; 7; 25

Step 3:
From top left, left to right: 15; 10; 35; 40; 31; 4; 33; 18

Test 45
Step 1:
From top left, left to right: 104; 45; 103; 69; 312; 23; 32; 192

Step 2:
From top left, left to right: 123; 234; 94; 144; 129; 159; 81; 19; 107; 140; 55; 54; 89; 69; 56

Step 3:
From top left, left to right: 26; 108; 25; 32; 111; 93; 84; 81; 100

Test 46
Step 1:
From top left, left to right: 100; 110; 110; 302; 291; 21; 63; 85

Step 2:
From top left, left to right: 250; 52; 73; 255; 210; 90; 58; 120; 301; 156; 222; 165; 111; 152; 476

Step 3:
From top left, left to right: 147; 70; 468; 210; 297; 624; 107; 8

Test 47
Step 1:
From top left, left to right: 212; 222; 192; 176; 427; 380; 234; 448

Step 2:
2. 522, 540
3. 752, 720
4. 1268, 1200
5. 1368, 1500
6. 3740, 3500

Other suitable estimates are possible.

Answers

Step 3:
From top left, left to right: 168; 259; 424; 1704; 3468; 5312

Test 48
Step 1:
From top left, left to right: 234; 420; 462; 552; 783; 544; 651; 498

Step 2:
2. 504, 540
3. 744, 720
4. 1482, 1500
5. 3402, 3500
6. 5824, 5600

Other suitable estimates are possible.

Step 3:
From top left, left to right: 354; 584; 792; 1432; 4068; 7152

Test 49
Step 1:
1. 480, 8, 60
2. 300, 6, 50
3. 210, 15, 14
4. 960, 16, 60

Step 2:
From top left, left to right: 96; 19; 40; 10; 5; 180; 2580; 125; 16; 15; 50; 24; 24; 18; 15

Step 3:
From top left, left to right: 20; 60; 5; 625; 15; 40; 19; 224

Test 50
Step 1:
From top left, left to right: 1040; 22; 26; 600; 120; 107; 56; 240; 65; 1920

Step 2:
From top left, left to right: 352; 35; 12; 500; 13; 5; 36; 360; 250; 253; 24; 5; 120; 40; 20

Step 3:
From top: 1300; 20; 24; 5; 3000; 21; 8; 392

Test 51
Step 1:
From top left, left to right: $\frac{1}{2}$; $\frac{1}{3}$; $\frac{1}{4}$; $\frac{1}{2}$; $\frac{1}{4}$

Step 2:
From top left, left to right: 6; 5; 5; 5; 2; 4; 4; 4; 8

Step 3:
1. F
2. F
3. T
4. F
5. F
6. F
7. T
8. F

Test 52
All answers are given in their simplest form. Equivalent answers are possible.

Step 1:
From top left, left to right: $\frac{1}{2}$; $\frac{1}{3}$; $\frac{1}{4}$; $\frac{1}{4}$; $\frac{1}{2}$; $\frac{3}{4}$; $\frac{2}{5}$; $\frac{7}{10}$; $\frac{3}{10}$

Step 2:
From top left, left to right: 4; 3; 10; 12; 18; 12; 6; 25; 3; 27

Step 3:
From top left, left to right: $\frac{3}{8}$; $\frac{1}{4}$; $\frac{3}{4}$; $\frac{3}{8}$; $\frac{1}{2}$; $\frac{1}{2}$

Test 53
Equivalent answers are possible.

Step 1:
From top left, left to right: $\frac{3+2}{7} = \frac{5}{7}$; $\frac{1+3}{5} = \frac{4}{5}$; $\frac{5+4}{10} = \frac{9}{10}$; $\frac{13+11}{25} = \frac{24}{25}$; $\frac{2+5}{8} = \frac{7}{8}$; $\frac{28+49}{123} = \frac{77}{123}$

Step 2:
From top left, left to right: $\frac{7}{9}$; $\frac{39}{58}$; $\frac{3}{4}$; $\frac{8}{11}$; $\frac{3}{4}$; $\frac{16}{19}$; $\frac{2}{3}$; $\frac{15}{17}$; $\frac{37}{47}$; $\frac{7}{8}$; $\frac{37}{39}$; $\frac{13}{20}$; $\frac{77}{108}$; $\frac{40}{209}$; $\frac{85}{107}$

Step 3:
From top left, left to right: $\frac{9}{10}$; $\frac{15}{17}$; $\frac{101}{137}$; $\frac{19}{29}$; $\frac{4}{18}$; $\frac{1}{24}$; $\frac{28}{225}$; $\frac{11}{108}$

Test 54
Equivalent answers are possible.

Step 1:
From top left, left to right: $\frac{3-2}{7} = \frac{1}{7}$; $\frac{3-1}{5} = \frac{2}{5}$; $\frac{5-4}{10} = \frac{1}{10}$; $\frac{5-2}{8} = \frac{3}{8}$; $\frac{13-8}{19} = \frac{5}{19}$; $\frac{11-3}{14} = \frac{4}{7}$; $\frac{15-7}{18} = \frac{4}{9}$; $\frac{28-12}{29} = \frac{16}{29}$; $\frac{13-7}{16} = \frac{3}{8}$

Step 2:
From top left, left to right: $\frac{4}{7}$; $\frac{1}{2}$; $\frac{8}{51}$; $\frac{1}{5}$; $\frac{17}{47}$; $\frac{4}{9}$; $\frac{7}{15}$; $\frac{5}{11}$; $\frac{1}{7}$; $\frac{4}{11}$; $\frac{3}{16}$; $\frac{1}{3}$; $\frac{44}{107}$; $\frac{15}{38}$; $\frac{38}{83}$

Step 3:
From top left, left to right: $\frac{2}{7}$; $\frac{12}{38}$ or $\frac{6}{19}$; $\frac{51}{104}$; $\frac{23}{47}$; $\frac{9}{100}$; $\frac{9}{45}$; $\frac{29}{64}$; $\frac{41}{72}$

Test 55
All answers are given in their simplest form. Equivalent answers are possible.

Step 1:
From top left, left to right: $\frac{5}{6}$; 1; $\frac{15}{109}$; $\frac{34}{45}$; $\frac{16}{71}$; $\frac{10}{19}$; $\frac{7}{24}$; $\frac{29}{37}$; $\frac{17}{32}$

Answers

Step 2:

From top left, left to right: $\frac{17}{19}$; $\frac{21}{48}$; $\frac{1}{4}$; $\frac{17}{37}$; $\frac{19}{35}$; $\frac{5}{6}$; 1; $\frac{18}{67}$; 1; $\frac{39}{56}$; $\frac{5}{13}$; $\frac{1}{9}$; 1; $\frac{3}{5}$; $\frac{5}{6}$; 1

Step 3:

From top left, left to right: $\frac{8}{15}$; $\frac{1}{2}$; $\frac{17}{39}$; $\frac{5}{6}$; $\frac{7}{12}$; $\frac{17}{27}$; $\frac{24}{47}$; $\frac{17}{31}$

Test 56

Equivalent answers are possible.

Step 1:

From top left, left to right: $\frac{9}{11}$; 1; $\frac{5}{6}$; $\frac{15}{19}$; $\frac{29}{31}$; $\frac{3}{20}$; $\frac{14}{41}$; 0; $\frac{1}{2}$

Step 2:

From top left, left to right: $\frac{15}{16}$; $\frac{11}{40}$; 1; $\frac{10}{13}$; $\frac{25}{26}$; 1; $\frac{17}{45}$; 1; $\frac{15}{37}$; $\frac{8}{37}$; $\frac{13}{30}$; $\frac{25}{74}$; $\frac{31}{47}$; $\frac{8}{29}$; $\frac{15}{43}$; $\frac{7}{18}$

Step 3:

From top left, left to right: $1\frac{1}{16}$; $\frac{73}{91}$; $\frac{19}{45}$; $\frac{61}{93}$; $\frac{23}{45}$; $\frac{40}{96}$; $\frac{7}{84}$; $\frac{23}{44}$

Test 57

Step 1:

From top left, left to right: 40; 33; 40; 210; 110; 50; 150; 60; 50; 70

Step 2:

From top left, left to right: 60; 60; 44; 180; 36; 100; 21; 48; 66; 105; 60; 90

Step 3:

From top left, left to right: 18; 42; 40; 66; 42; 70; 72; 55

Test 58

Step 1:

Circled: 0.15; 7.23; 100.04; 0.3; 20.1; 0.04; 30.2

Step 2:

From top, left to right: 1, 50, one point five (zero); 4, 25, four point two five; 1, 20, one point two (zero); 2, 18, two point one eight; 49, 90, forty-nine point nine (zero)

Step 3:

From top: 2200; 7600; 4, 58; 85; 1600; 900

Test 59

Step 1:

Equivalent fractions are possible:

1. $\frac{7}{10}$, 0.7
2. $\frac{6}{10}$, 0.6
3. $\frac{4}{10}$, 0.4
4. $\frac{43}{100}$, 0.43

Step 2:

1. From top left, left to right: $\frac{3}{10}$; $\frac{6}{10}$; $\frac{9}{10}$; 0.1; 0.5; 0.9

2. From top left, left to right: $\frac{3}{100}$; $\frac{5}{100}$; $\frac{14}{100}$; 0.05; 0.08; 0.11; 0.18

Step 3:

From top left, left to right: 0.29; 0.6; 0.718; 0.805; 0.1; 0.9; 0.143; 0.12

Test 60

Step 1:

From top left, left to right: $\frac{1}{10}$; $\frac{7}{10}$; $\frac{8}{10}$; 0.3; 0.5; 0.8

Step 2:

1. 0.2; 0.15; 0.24; 1.36; 0.97; 0.08

2. Equivalent answers are possible:

$\frac{6}{10}$; $\frac{28}{100}$; $\frac{4}{100}$; $\frac{128}{100}$; $\frac{7}{100}$

Step 3:

From top left, left to right: 0.3; 0.18; 0.05; 0.09; 0.12; 0.4; 0.91; 2.7

Test 61

Step 1:

From top left, left to right: 0.8; 0.72; 0.36; 0.81; 1.93; $\frac{7}{10}$; $\frac{57}{100}$; $\frac{2}{100}$; $\frac{37}{10}$; $\frac{35}{100}$

Step 2:

1. 0.1, 0.2, 0.5, 1
2. 0.01, 0.05, 0.1
3. 0.23, 2.3
4. 1, 10
5. 0.99, 9.9
6. 3.6
7. Ninety-two, 0.1s

Step 3:

1. 5, 3
2. 53, 9, 82
3. 7, 39, 77
4. 9, 90
5. 2.1

Test 62

Step 1:

From top left, left to right: 0.25; 31.31; 100.7; 6.54; 7.42; 63.18; 0.05; 40.09

Step 2:

1. 1, 0, 8, 108
2. 6, 1, 5, 2, 6152
3. 5, 7, 57
4. 3.1

Answers

5. 485.96
6. 8, 80
7. 0.25
8. 3

Step 3:
From top: nine 10s, eight 1s, six 0.1s and eight 0.01s; eight 10s, nine 1s, one 0.1 and two 0.01s; one 100, seven 0.1s and eight 0.01s; seven 0.1s and two 0.01s; seven 1s and five 0.01s
Other suitable answers are possible.

Test 63
Step 1:
From top left, left to right: zero point eight five → 0.85; thirty-four point zero eight → 34.08; zero point zero five → 0.05; thirty-four point eight → 34.8

Step 2:
Left table: one point zero five; one hundred point zero one; three point five two; zero point zero two
Right table: zero point zero one; fifty-one point four five; nine point zero two; eight point zero seven; zero point zero six

Step 3:
From top left, left to right: 90.5; 0.13; 20.80; 0.02; 0.87; 6.09; 12.07; 10.01

Test 64
Step 1:
From top left, left to right: $\frac{5}{100}$ or $\frac{1}{20}$, 0.05; $\frac{70}{100}$ or $\frac{7}{10}$, 0.7; $\frac{29}{100}$, 0.29

Step 2:
Left column: 101.01; 25.34; 382.03; 3.04; 1200.05
Right column: 51.15; 0.05; 1.02; 10.06; 383.0

Step 3:
From top: one hundred and three point one five; zero point zero five; eight point zero zero; one thousand and six point one five; seventy point zero nine; forty point three eight; forty-two point one nine; seven point zero nine

Test 65
Step 1:
From top left, left to right: >; <; >; <; <; >; >; <

Step 2:
From top left, left to right: >; >; >; >; =; >; <; >; =; <; =; <; >; >; <; >; >; =

Step 3:
From top left, left to right: >; <; <; <; <; =; =; <; >; =

Test 66
Step 1:
From top left, left to right: <; >; >; >; >; <; >; <

Step 2:
From top left, left to right: <; <; <; <; >; >; >; <; <; <; <; >; <; <; >; =; <; >

Step 3:
From top left, left to right: >; =; <; <; <; =; <; <; <; <

Test 67
Step 1:
From top left, left to right: 0.7; 0.03; 0.38; 3.92; 0.09; 0.21; 0.7; 0.05; 0.52; 5.42

Step 2:
From top left, left to right: <; <; <; <; >; =; <; <; <; =; >; <; >; <; <; =

Step 3:
1. 5.0, 5.05, 5.5, 5.55, 50
2. 0.33, 3, 3.03, 3.3, 33

Test 68
Step 1:
From top left, left to right: 2.1; 70; 0.02; 90; 7.04; 54.01; 17.3; 0.08; 0.7; 9.23

Step 2:
1. $\frac{3}{10}$, 0.3
2. $\frac{35}{1000}$, 0.035
3. $2\frac{1}{4}$, 2.25
4. tenths; hundredths
5. 0.49, 0.51
6. 1.99, 2.01

Step 3:
From top left, left to right: >; <; =; >; >; <; =; <

Test 69
Step 1:
Right angles: 1, 3, 5, 7
Acute angles: 8, 9
Obtuse angles: 2, 4, 6, 10

Step 2:
Right angles: 3:00, 9:00
Acute angles: 10:00, 7:45, 5:30, 2:00
Obtuse angles: 9:30, 1:30, 11:30

Step 3:
∠ 2: obtuse; ∠ 3: right; ∠ 4: right;
∠ 5: right; ∠ 6: right; ∠ 7: obtuse

Answers

Test 70
Step 1:
1. line segments, triangle
2. four
3. line segments, pentagon

Step 2:
Triangles: 4
Quadrilaterals: 1, 3, 8, 9
Pentagons: 5, 7
Other: 2, 6

Step 3:
From top left, left to right:
3, 3, triangle;
4, 4, quadrilateral;
4, 4, quadrilateral;
5, 5, pentagon

Test 71
Step 1:
From left to right:
1. acute, right-angled, obtuse
2. obtuse
3. acute
4. right-angled
5. scalene, isosceles, equilateral

Step 2:
Acute: 1, 2, 3, 4, 6, 10
Obtuse: 7, 9
Right-angled: 5, 8
Isosceles: 1, 3, 4, 5, 9
Equilateral: 1, 4

Step 3:
From left to right:
1. 2
2. isosceles
3. isosceles, equilateral
4. sides, angles

Test 72
Step 1:
Shapes 1, 2, 3 and 5 ticked and lines of symmetry drawn

1.
2.
3.
5.

Step 2:
Shapes 2, 3, 4 and 6 ticked and lines of symmetry drawn

2.
3.
4.
6.

Step 3:
1. 2; 4
2. 3
3. 1
4. infinite

Test 73
Step 1:
1. length, width; 15
2. side length, side length; 16

Step 2:
From top left, left to right:
1. 35; 300; 1750; 2940; 2700; 800
2. 225; 6400; 121; 400

Step 3:
From top left, left to right:
1. 35 000; 3200; 8000
2. 250 000; 490 000; 160 000

Test 74
Step 1:
1. m, m, 40 m²
2. cm, cm, 20 000 cm²
3. cm, cm, 60 cm²
4. m, m, 5000 m²

Step 2:
From top left, left to right:
1. 1040; 20; 160; 480; 1350; 420
2. 144; 8100; 256; 1600

Step 3:
From top:
1. 12; 40; 16
2. 70; 90; 12

Answers

Test 75

Step 1:
From left to right: 15; 16; 10; 49

Step 2:
From left to right:
First table: 32 cm^2, 360 m^2, 20 cm, 24 cm, 18 m
Second table: 81 cm^2, 196 m^2, 6 cm, 10 cm, 14 cm

Step 3:
From top left, left to right:
1. 10; 60; 12; 32
2. 15; 20; 12; 18

Test 76

Step 1:
1. m
2. km
3. m
4. km
5. m
6. km
7. km
8. m

Step 2:
From top left, left to right: 1000; 4000; 30; 9000; 5; 18; 7; 80 000; 38; 19 000; 8750; 7007; 99; 16 040; 3450

Step 3:
From top left, left to right: 5020; 9067; 4300; 7800; 7570; 7920; 29 400; 2

Test 77

Step 1:
1. perimeter
2. 2 × length + 2 × width (in either order)
3. 4 × side length
4. From left: 25, 26, 26; 28

Step 2:
From left to right:
First table: 56 cm, 90 cm, 150 cm, 184 cm, 360 cm
Second table: 136 cm, 200 cm, 288 cm, 352 cm, 172 cm

Step 3:
From top:
1. 16; 26; 12; 32
2. 19; 23; 16; 12

Test 78

Step 1:
1. From top: 22; 40; 82; 156
2. From top left, left to right: 96; 72; 240; 376

Step 2:
From left to right:
First table: 15 cm, 40 cm, 46 m, 5 m, 11 m
Second table: 92 cm, 18 m, 8 m, 60 cm, 24 m

Step 3:
1. 160
2. 340
3. 28
4. 144

Test 79

Step 1:
From top left, left to right: 60; 60; 24; 7; 12; 3600; 100; 120

Step 2:
From top left, left to right: 240; 210; 1.25; 36; 24; 16; 6; 80; 950; 0.70; 14; 6.08

Step 3:
From top left, left to right: 11, 16; 5, 90; 36, 2160; 0.75, 2700; 28, 672; 30

Test 80

Step 1:
From top left, left to right: 15; 232; 240; 32; 693; 106; 761; 1480

Step 2:
From top left, left to right: (930 − 690) ÷ 5 = 48;
523 − (609 ÷ 3) = 320; (36 × 23) + 123 = 951;
1008 − 41 × (15 + 8) = 65; (979 − 115) ÷ (3 + 13) = 54

Step 3:
From top left, left to right: 87; 51; 1050; 198; 56; 120; 40; 630; 1984

Notes